MW01591418

SECRET WEAPON

U.S. High-Frequency Direction Finding in the Battle of the Atlantic

Kathleen Broome Williams

Naval Institute Press
Annapolis, Maryland

Library of Congress Cataloging-in-Publication Data
Williams, Kathleen Broome, 1944–
  Secret weapon : U.S. high-frequency direction finding in the
Battle of the Atlantic / Kathleen Broome Williams.
      p.   cm.
  Includes bibliographical references and index.
  ISBN 1-55750-935-2 (alk. paper)
  1. Radio direction finders. 2. World War, 1939–1945—Campaigns—
Atlantic Ocean. 3. World War, 1939–1945—Naval operations,
American. I. Title.
TK6565.D5W55   1996
940.54'516—dc20   96-14583

Printed in the United States of America on acid-free paper ∞
03 02 01 00 99 98 97 96   9 8 7 6 5 4 3 2
First printing

In memory of my father

Contents

Illustrations

Foreword

The trump of the submarine was, and still is, its invisibility when submerged. With the development between World Wars I and II of asdic or sonar, many in the 1930s thought that the danger from U-boats had been overcome. But in 1939–40, when the German U-boats attacked convoys in surfaced "wolf packs" at night, sonar could not help. The British at first made great efforts to equip escort vessels with radar, in an attempt to turn the night into day.

For many years after World War II, most writers, historians, and even those who experienced it considered radar the single most important piece of equipment in the fight against the German U-boats. They ascribed the decisive defeat of the U-boats in the May 1943 Battle of the Atlantic mostly to radar, because it gave the escorts and aircraft "sight," allowing them to attack at night and even in fog.

In the 1960s it became possible to study convoy battles in detail, by comparing U-boat war diaries with Allied escorts' action reports. Slowly it became apparent that there was another instrument in the "war in the ether," which in 1942–43 was in many cases more important than radar: the high-frequency direction finder (HF/DF) aboard the escort vessels. But in the mid-1970s, discoveries about Ultra overshadowed HF/DF. The decryption of U-boat radio signals by the British cryptanalysts at Bletchley Park was finally established as the main reason for the successful rerouting of the convoys around the German wolf packs.

The role played by HF/DF was nonetheless important, and Kathleen Williams now restores it to its rightful place in history. Her careful tracing in particular of the French and American parts of HF/DF's development offers new insights into the Battle of the Atlantic.

Captain Karl Dönitz was officer in command of the new U-boat force between 1935 and 1939. After considerable thought, experiments, and exercises, he decided on his so-called group or wolf-pack tactics to use against the expected Allied convoy system. Part of his concept was that it might be possible to control U-boat deployments by radio communications between the shore command, tactical leaders, and individual

U-boats at sea. Using a short-signal code, he hoped to make the signals so short that direction finders (D/F) would not have enough time to take an exact bearing. He was strengthened in this belief during the first year of the war, when the German *B-Dienst* decrypted several French radio messages giving positions of U-boats from D/F that were far from the real positions of the signaling boats. Further encouragement came when U-boats, ordered to send regular weather reports from positions far out in the Atlantic, were overrun by convoys.

So the fear of D/F locating the U-boats dwindled. German experts thought HF/DF could not be used aboard small warships like convoy escorts because of their space requirements and their weight, thus nobody saw any direct danger for the U-boats when they sent their contact signals during convoy operations. At those moments, the enemy did know that there were U-boats in contact.

In fact, it would become possible to reduce the space and weight requirements of a HF/DF set. Further, French inventors Busignies and Deloraine developed a method to combine the Adcock aerial with the cathode-ray tube, achieving an instant bearing. The failure of German experts to predict any of this, or the accomplishments of the British "father of radar," Sir Robert Watson-Watt, also in the field of HF/DF, had grave consequences for U-boats.

Attempts to produce a short-wave D/F set for smaller ships, using old D/F methods, had in Britain led to a prototype FH.1 model, installed on 12 March 1940 on the destroyer *Hesperus*. But it was unsatisfactory. In July 1941 an improved set, FH.2, was ready, and a little later the FH.3, developed finally in accordance with Watson-Watt's suggestions. But the prototype aboard the sloop *Culver* was lost on 31 January 1942, when U-105 sank the sloop while it was escorting the convoy SL.93. At that time, the convoy's rescue ships were equipped with the first FH.3 sets.

On 21 February 1942, when U-155 had sent a contact signal from convoy ON.67, the rescue ship *Toward* took a bearing of this signal. The commander of the U.S. escort group A.6 sent the destroyer *Lea* out along the bearing line, but U-155 dived before the destroyer had won contact.

The following incident ended in failure as well, when from 11 to 13 May the six U-boats of group Hecht attacked the convoy ONS.92 and sank seven ships. The American escort commander of group A.3, Commander Heffernan, had failed to understand the many HF/DF bearings the rescue ship *Bury* reported.

Such first experiences might have led to some reluctance in equipping escorts with the new HF/DF sets. They brought additional top weight to the already installed radar sets, which most of the commanding officers regarded as more effective in locating U-boats.

But some other captains anticipated the capabilities of the HF/DF set. The captain of the Canadian destroyer *Restigouche,* on his own initiative during a refitting in the winter of 1942 in Britain, succeeded in obtaining an FH.3 set. When his ship joined the escort group A.3 of convoy ONS.102, now led by Cdr. P. R. Heineman, U.S. Navy, the *Restigouche's* HF/DF operator was able to take bearings of most of the contact signals so that escorts could run them down, forcing the U-boats to dive.

One of these, U-94 under Lieutenant Ites, was attacked with depth charges and damaged. Ites made it back to port and reported that he assumed to have been D/F'd when sending his contact signal and then run down. He was advised by the staff experts that "this must have been *Funkmess,*" as radar was called in Germany.

The fear of radar was so much in the forefront of all German thinking then, and almost until the end of the war, that nobody took note of the possibility of HF/DF sets aboard the escorts. This was despite the fact, that in early 1943, the German *B-Dienst* decrypted many signals to and from convoys and escort groups, signals that indicated possible HF/DF use. On 9 April 1943, a signal was decrypted showing that the flagship of the escort group TU.24.1.9, USCGC *Spencer* (now the flagship of Commander Heineman), was equipped with a HF/DF set. Further, German agents opposite Gibraltar, located in a building at Algeciras, sent photographs of destroyers—HMS *Volunteer* or HMS *Antelope*— clearly showing masts with the Adcock aerial on top.

But neither the technical experts nor the U-boat command saw the originals of the photos. They were sent to the ship-identification service, where they were retouched to disguise the source. Then they were printed in thousands and sent to the commands and ships of the fleet. But the aerial masts had been retouched to cover up the Gibraltar houses in the background!

This mismanagement in the evaluation of available intelligence, a consequence mainly of the fixation on radar, prevented the realization that many convoys escaped the wolf packs. The first contact signal was D/F'd, and the U-boat was forced to dive while the convoy took an evasive turn so that the contact was lost. Of 174 North Atlantic convoys between July 1942 and May 1943, 105 had no U-boat contact because they were rerouted or passed areas with no U-boats. But of the 69 that were reported at least once by U-boats, 23 escaped without any loss and 30 sustained only minor losses. We now know that this was to a great extent because escorts or aircraft running down HF/DF bearings forced U-boats to dive or to turn away.

Beginning in the fall of 1943, escorts running down HF/DF bearings came into promising firing positions for the new acoustic homing torpedoes *Zaunkönig* ("Gnat"). The first hit was achieved by U-270 against the frigate *Lagan* of convoy ON.202, on 20 September 1943. But such situations were not fully exploited, because of the rigidity of the German experts and the command's obsession with radar and the need to use radio messages to conduct operations.

It took many months before the British escorts got their effective HF/DF sets, and Canadian ships were always at the end of the line with new equipment. But the American side also had its problems, and this book describes them vividly.

The introduction of HF/DF into the fleet gave rise to bureaucratic infighting between various technical offices and their heads, who did not want to use foreigners' inventions. It is of great interest to see how some strong personalities outside the U.S. Navy forced the development through bottlenecks in a surprisingly short time; the first destroyer, *Corry*, got its test set as early as May 1942. A shore HF/DF station was responsible for the first real success, when a Mariner flying boat of Patrol Squadron 74 ran down the bearing of a long signal of U-158. Relaying information about its operations and successes in the Gulf of Mexico, this U-boat was surprised and sunk in the area of Bermuda on 30 June 1942.

It was the beginning of great successes the American hunter-killer groups achieved from 1943 to 1945, not only with the help of Ultra but also, to a high degree, by using their HF/DF sets. Kathleen Williams's fine work represents an important contribution toward our understanding of these advances in technological warfare.

—*Jürgen Rohwer*

Preface

The Battle of the Atlantic was a bitter, highly complex, and long, drawn-out struggle. British survival depended on a flow of goods crossing the ocean in thousands of Allied and neutral cargo ships. German victory depended on their submarines sinking more cargo ships than could be replaced. The consequences of failure for both sides were severe, because in many ways the outcome of World War II in Europe hinged on the naval campaign in the Atlantic.

Britain played a crucial role in the European war—as a source of vital supplies for the Soviet Union, as a base of support for Allied operations in the Mediterranean, and as a staging area for the invasion of France. By the spring of 1943, however, the German U-boats had come very close to forcing Britain out of the conflict.

They sank 108 ships in March alone, reaching a pinnacle of success. Just at that point, though, all the Allied antisubmarine warfare measures began to come together effectively, and by May the tide was turning against the U-boats.

Much has been written about the multiplicity of factors that brought victory to the Allies in the longest battle of World War II. Convoying, more and better convoy-escort vessels, escort carriers with support and hunter-killer groups, air coverage, new weapons, new tactics, and the eventual flood of new American ships, each played a significant and complementary part.

But perhaps most important of all in combating the U-boat threat was the successful Allied use of radio intelligence. Once the British broke the German naval radio codes and ciphers, they could periodically obtain information (referred to as Ultra) that could be used to route convoys safely around gathering U-boat wolf-packs. By making possible these evasive maneuvers, the information garnered from intercepted and decrypted German radio messages may have done more to save merchantmen from attack than any other single factor in the Atlantic campaign. But Ultra was neither always available nor timely. Often the decryption process took too long for the resulting information to be of operational use.

Ultra was only one way of locating U-boats in the Atlantic, however; other successful devices had emerged from interwar developments in electronics. This was the first truly electronic war, and sonar, radar, and shipborne direction finders each provided the Allies with short-range, tactical U-boat location capabilities. Land-based direction finders—long-range detection devices—produced information that, like Ultra, could be used in the diversion of convoys. While sonar and radar are well-known technologies, Huff Duff (high-frequency direction finding, or HF/DF) has remained relatively obscure.

HF/DF was a radio electronic device used extensively, often in conjunction with sonar and radar, to locate German U-boats. It has been given much credit for helping the Allies to win the close-run contest in the Atlantic. In his official history of the U.S. Navy in World War II, Samuel Eliot Morison described HF/DF as a convoy's "highly sensitive and elongated cat's whiskers."[1] Sir Arthur Hezlet, a British authority on naval electronics, wrote that during the Atlantic campaign HF/DF became "an important anti-submarine device."[2]

In spite of the significance of electronics to modern naval warfare and to the Battle of the Atlantic in particular, HF/DF has been covered less extensively than have radar, sonar, and especially Ultra. Recently there have been some vivid accounts of its tactical use,[3] but in general the record of the development of HF/DF has been confined mostly to technical journals. My objective in this book is to remedy one part of the deficiency by examining the history of American naval HF/DF, particularly shipborne HF/DF.

Like sonar and radar, HF/DF was developed in the 1920s and 1930s in several different countries at the same time. Direction finders intercept radio waves and indicate their direction of arrival. The desire to locate U-boats by their radio signals spurred developments in the direction-finding field, and HF/DF evolved as part of the whole spectrum of electronic countermeasures.

The British and the Canadians carried the major burden of the campaign against the U-boats in the North Atlantic; the British were also the first to use HF/DF effectively. By mid-1942 they had developed roughly accurate shipboard HF/DF sets, which were produced and installed on Royal Navy vessels escorting merchant convoys. At that point HF/DF began to reach its full power as a weapon of tactical offense, enabling escorts to locate, attack, and sink U-boats lurking near the convoys.

There was no airborne HF/DF (unlike radar), because the antennas were too large to fit on aircraft. Allied air forces did eventually make good use of HF/DF, but this was shore based, often in mobile units, and was used for navigation.

In the United States a practical HF/DF device was developed and was brought to the attention of the U.S. Navy. Though American naval HF/DF appeared later than its British cousin, contrary to popular perception it was neither an adaptation of, nor did it derive from, the British device. Indeed, the American shipborne apparatus finally put to use in the Atlantic campaign had been available to the U.S. Navy before the British had deployed an effective device. It was also based on a more advanced technology than that of the initial and most pervasive British sets.

The origins and development of British naval HF/DF are beyond the scope of this book, but they clearly remain an important area for future investigation. I have concentrated on unraveling the genesis of the American device, and on suggesting the nature and significance of its role.

HF/DF did not emerge from an official plan for naval electronic development. Instead it was an example of development from the bottom up; in many ways fortuitous, unexpected, and in some quarters unwanted. The HF/DF devices deployed by the U.S. Navy were the creation of Henri Busignies, a French engineer at ITT in Paris. Aggressive marketing by Sosthenes Behn, the president of ITT, introduced the strategic, land-based device to top American military at the beginning of 1941. But though Busignies also had an apparatus ready that was designed for use at sea, the need for tactical shipborne equipment was not recognized by navy decision makers until spring 1942.

At that point a change in policy was inspired in large part by the example of British tactical successes with shipborne HF/DF. The evident superiority of Busignies's device, successfully promoted by Behn, ensured its adoption. Thus by mid-1943 the U.S. Navy was ready with its own shipborne HF/DF, and used it to good effect against the U-boats in its Central Atlantic area of responsibility.

After identifying its electronic-technology needs, the U.S. Navy decided on particular devices and then arranged for their production, procurement, and supply. Finally, shipborne HF/DF had to be installed onboard the ships and the technicians trained before the technology could be used in combat. There was never a rigid system controlling development, however, which the independent evolution of HF/DF illustrates well.

The Allies won the Battle of the Atlantic thanks in part to the use of HF/DF, and continuing postwar developments indicate its lasting significance as an antisubmarine device. But there was no single panacea, no miracle weapon, device, or even intelligence breakthrough that would win the Atlantic campaign for the Allies. HF/DF was one element only, though an important one. It took the combined weight of them all to defeat Dönitz's skilled and dedicated submariners.

Summing up a recent conference on the Battle of the Atlantic, Geoffrey Till wrote that "we historians should hope to find no more than multiple and approximate causes" for the Allied victory.[4] This book does not give a comprehensive accounting, nor even a measured overview, of all the causes. Instead, it focuses on American HF/DF, one hitherto neglected cause of victory, as it was developed by one of the Allies.

✦ ✦ ✦

This book would not have been written without the encouragement and help of a number of very special people. Above all, I am grateful to Mme Cécile Busignies and to Mrs. Ruth Lockhart Lombardi for their generosity with documents, time, and memories. In many ways this book is due to them. Mrs. Barbara Goldstein Koz Paley has kindly allowed me to consult her father's papers. I should also like to thank Prof. David Syrett, who suggested Huff Duff to me; Prof. Patrick Abbazia; Prof. Jürgen Rohwer; and Prof. George Phillips. Each read various versions of the work and made many valuable suggestions.

I am very much obliged to Bernard Cavalcante, Kathleen Lloyd, and Regina Akers at the Naval Historical Center; Dr. David van Keuren, Murray Bradley, Dean Bundy, and Jim Lucas at the Naval Research Laboratory; Scott Price at the Historian's Office, U.S. Coast Guard; and archivists at the National Archives. Tony Simpson and the staff of John Jay College of Criminal Justice Library, CUNY, have been extremely helpful, as have librarians at CUNY's Graduate Center, and Pat Root and her colleagues at the Bronxville Library. The encouragement of my editor, Anne Collier Rehill, long kept me going.

My stepfather, George Plante, was a wireless operator and commissioned war artist in the British merchant navy. He was twice torpedoed on the North Atlantic run, the second time during the famous convoy battles of March 1943. His tales of those adventures gave me an early interest in naval warfare, and his paintings and sketches brought the past even more vividly alive. He was an enthusiastic supporter of and contributor to this book, and I deeply regret that he did not live to see it in print.

I also extend my thanks to all those others who patiently answered my questions about their wartime experiences, especially Eugene Lombardi, Capt. Ben Brooks, Karl Degener-Böning, Avery Richardson, Adm. Frederick Furth, Wolfgang Hirschfeld, and Bob Seroskie. Others have contributed in ways too numerous to list—for their encouragement and help at every stage of this endeavor I am greatly indebted to Capt. Scott Lothrop; Dr. Norman Friedman; Milan Gupta; Richard DiNardo; Albert Nofi; Wayne Sarf; John Dikeos; Roger Broome V; my mother, Jane Plante; and my children, Brooke, Alexandra, and Tara.

With so much assistance, I have been saved from numerous errors of omission and commission. Those errors that remain are entirely my own.

CHAPTER 1

An Overview

It was a scene repeated thousands of times during the Battle of the Atlantic. Early on the morning of 17 March 1943, the British merchant tanker *Southern Princess,* loaded with ten thousand tons of fuel oil, was nine days out of New York making its way with convoy HX 229, bound for Britain. Forty-eight years later, the ship's second wireless officer, George Plante, remembered the chaos as though it had just happened.

> Ships were being picked off. An explosion would shatter the silence and flames lighten the night. Some were far, some were near. There was no pattern. Sometimes we could see which ship it was from its position in the convoy; much of the time we just had to take a guess. Cargo ships burned and tankers exploded. The most spectacular displays happened when a benzene tanker caught it.
>
> I had relieved the first officer for a few minutes and was alone in the wireless cabin when the torpedo struck. The force of the explosion threw me across the cabin and left me in a heap on the deck. It has been reported that loud, agonizing screams were heard coming from my ship but I didn't hear any screams. Missing the last lifeboat when I returned to the cabin to get the emergency wireless transmitter, I took my shoes off, neatly arranged them on the deck, and jumped. It isn't difficult to jump into the middle of the Atlantic at three o'clock in the morning when flames are attacking your rear.[1]

The *Southern Princess* sank in a fiery inferno. It was a sad coincidence that this ship, previously named the *San Patricio,* should be destroyed on Saint Patrick's Day. But at least the loss of life was not great, which was against all odds given the nature of the cargo. Wireless Officer Plante was pulled from the ocean, along with all but two of his fellow crewmen and

Sinking of the *Southern Princess.* Painted by commissioned war artist and second wireless officer George Plante.
Imperial War Museum

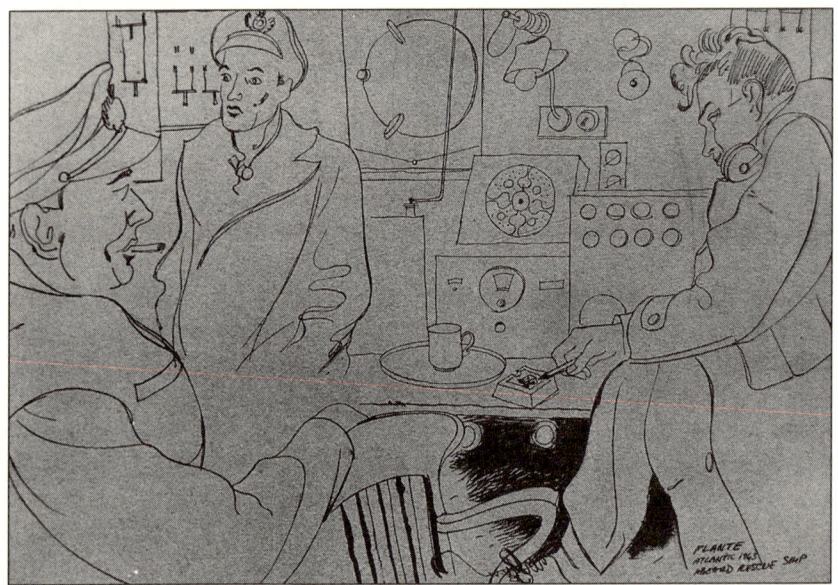

Radio cabin on the *Tekoa,* March 1943. Ink sketch by commissioned war artist and second wireless officer George Plante.
Courtesy of George Plante

two passengers, thanks to the courage of Capt. Albert Hocken of the New Zealand freighter *Tekoa* (which means "Good Luck" in Maori). In defiance of standing orders, the youthful Captain Hocken, on his first voyage as master, stopped his ship in the U-boat-infested waters to pick up survivors.

The fast convoy HX 229 had sailed from New York on 5 March in conjunction with slow convoy SC 122. They headed northeastward off the North American coast, clearing Cape Cod by 150 miles. The ships sailed in formation past Nova Scotia and Newfoundland, and then gathered themselves for the sprint across the open ocean to Britain. But the convoys were set upon in mid-Atlantic, caught in the "Air Gap" beyond the protection of friendly planes. They fell victim to one of the most successful concentrations of German U-boat "wolf packs" in the war, and before they sighted land again off Northern Ireland on 22 March, the two convoys suffered severe casualties.

A total of thirty-eight U-boats assembled in three groups along the path of the ninety merchant ships. The Germans sank twenty-one ships, including the *Southern Princess,* representing 20 percent of the vessels involved. Only one U-boat was lost.[2]

The battle for convoys HX 229 and SC 122, which took place between 16 and 20 March 1943, is often considered the climactic convoy battle of World War II. It was one of the biggest convoy operations of the war, and at the time it seemed to confirm the U-boats' domination of Allied supply lines to the British Isles. But the victory was hollow: during the next two months there was a dramatic drop in the number of merchant-ship sinkings. In fact, April and May 1943 were two of the least successful months of the whole U-boat campaign, and on 24 May the submarines were withdrawn from the North Atlantic and sent in search of easier targets to the south.[3]

This move marked the turning point in the Atlantic War. From then on the hunters, those sleek and dangerous gray wolves, became the prey until, finally, they were decisively defeated. This was a deadly contest between one weapon-delivery system, the U-boat, and a whole array developed to oppose it. The U-boat operated underwater and on the surface, but its opponents took to the air as well. And both sides filled the airwaves with an almost incessant flow of radio communications, orders, and instructions. Naturally, then, many vital elements were involved in the Allied victory over the U-boat. A whole range of Allied antisubmarine-warfare (ASW) measures, all increasingly well coordinated, finally came into play together. These included vessels, weapons, tactics, technical devices, and intelligence information.

Yet of all the elements available to both sides in the Atlantic campaign, "wireless was indisputably the single most important one."[4] In recent years, considerable attention has focused on the pivotal role of radio intelligence in defeating the submarine menace. As previously classified information has become available, an ever-sharper picture has emerged of the extensive Allied reliance on intelligence from breaking of German naval radio codes. This "special intelligence," as it was called by the British, has become well known as Ultra. The persistent German radio use that made codebreaking possible also made the U-boats vulnerable to Allied direction finding (D/F), and high-frequency direction finding (HF/DF), also known as Huff Duff, has frequently been cited as another vital element in the radio war against the U-boat.[5]

HF/DF is a way of locating a transmitter by ascertaining the direction of arrival of the radio waves it sends out. Every time a U-boat in the Atlantic sent a radio message, it exposed itself to direction finding; eventually direction finders on ships were used effectively to locate transmitting U-boats.

Relatively little has been written about the technical difficulties of developing a viable shipborne HF/DF, and still less about the way that development was inspired and came about. This work focuses on the American side of the question: where and how the U.S. Navy acquired an electronic device that it could use to locate German submarines by their radio transmissions. The search for answers sheds some light as well on the larger question of how a democracy came to arm itself with the appropriate technology to fight an unwanted war for which it had been grievously unprepared.

To put the development of American HF/DF in perspective, we need to consider what caused the costly confrontation at sea and what factors were involved in determining its course. Once before, during the First World War, German submarines had come perilously close to forcing a British surrender.

As an island nation, Britain relied for survival on imports brought in by sea. Early in 1917, in a desperate gamble to knock its enemy out of the war, Germany launched an unrestricted submarine attack on all vessels approaching the island, even neutrals. Germany knew this would almost certainly bring the United States into the war and hoped to starve Britain into surrender before U.S. troops could be mobilized, trained, and used on the western front.

At first the U-boats sank so many ships carrying vital supplies that they almost succeeded in cutting Britain's maritime umbilical cord. But in a last-ditch attempt to stave off further sinkings, the British resorted to the ancient trade-protection technique of gathering merchant ships into convoys. No one knew if convoying would be effective under the conditions of modern warfare and against submarines; some argued that it would just present the Germans with more visible and vulnerable concentrations of targets. But pressure from young staff officers in the British Admiralty, as well as the recommendations of U.S. Admiral W. S. Sims, persuaded the Royal Navy (RN) to give it a try. The results were impressive, and sinkings obtained by the German U-boats were quickly reduced.

Then, in the 1930s, the battleship fixation that mesmerized most interwar navies gripped Germany too, and a new *guerre de course,* or war against trade, was neither expected nor planned. Hitler did not view his navy as a war-winning weapon at all. For him the Atlantic campaign was initially "an unexpected maritime dimension" to the land war he wanted.[6]

Hitler had intended to win a war on the continent with his army, in a series of bold and quick strikes with the support of the German air force. The navy did not receive the funding it believed it needed to prosecute a war, and such resources as were authorized mostly went toward the prestigious surface fleet. It was not until September 1939 that the commander in chief of the navy, Grand Adm. Erich Raeder, finally abandoned the long-term plan to create a balanced fleet and concentrated instead on U-boat construction.

At the outbreak of war on 3 September, Capt. Karl Dönitz, head of the U-boat arm, had only thirty-nine operational U-boats with which to initiate what would become the longest campaign of World War II. The Battle of the Atlantic continued for sixty-eight months, ending only with Germany's surrender on 8 May 1945. At first Hitler put his naval forces under strict instructions to observe maritime rules of neutrality. For two frustrating years they were somewhat held in check to avoid repeating the error of World War I by antagonizing the United States once again.[7]

Only when it was finally clear to Hitler that he was locked in a long struggle of attrition with Britain was the navy able to compete more effectively for resources comparable to those of Germany's army and air force. It was June 1942 before Hitler belatedly came around to the view of Admirals Raeder and Dönitz that U-boats could defeat Britain and help win the war. Submarines finally began to get increased funding for repairs and above all for priority building programs.[8] This reluctant expedient almost won the war for Germany.

Meanwhile, in the intervening years since the Peace of Versailles, the British Isles had become more dependent than ever on goods from abroad, and so even more vulnerable to attack at sea. In 1939 Britain's total annual import requirements were 55 million tons. Although self-sufficient in coal, industrialized Britain imported all its oil and more than half of its foodstuffs. Britain also relied upon imports for most metal ores, all rubber, and even quantities of key manufactured goods.[9] Dependent upon overseas trade for many of the materials vital to a modern state, Britain was once again in danger of becoming easy prey for a massive submarine offensive.[10]

By spring 1941, after only a year and a half of war, a weak U-boat campaign had succeeded in cutting goods reaching Britain to an annualized rate of 28 million tons, and losses were still mounting. Food imports accounted for half of the cargoes, having reached "an irreducible minimum of 15 million tons." That rate forced Britons to a diet consisting mostly of bread, potatoes, and vegetables.[11] Since livestock had been slaughtered to cut back on the need for imported fodder, meat was very scarce. So were eggs; each person was allowed only one egg every two weeks. "Everyone was in the front line in one way or another," remembers one man; "sugar, chocolate, cheese, butter, meat, petrol, clothing, even such essentials as gin and scotch were strictly rationed! Everything was either in short supply or nonexistent."[12] By this time the Germans were sinking ships at twice the rate of new construction, and it was not at all clear that Britain could hold on for much longer.

As supplies dwindled, people prepared to fight off the inevitable cross-channel invasion. Then, unexpectedly, the fear of invasion lifted when Hitler launched Operation Barbarossa against Russia in June 1941. But still the U-boat war continued, and it became harder and harder for supply convoys to get through the blockade.

Between the wars the Royal Navy, like other navies, had been dominated by battleship men. They seemed to have forgotten that it was German U-boats, not battleships, that had almost brought Britannia down in 1917. As a result, in 1939 the Royal Navy was painfully deficient in convoy escorts and antisubmarine vessels and equipment of all sorts. Once war broke out, however, Britons quickly became keenly aware that they could not long survive if their naval forces failed to maintain the Atlantic Ocean as an unobstructed highway.[13]

This objective was also of importance to the neutral United States, for whom freedom of the seas was an economic—and for some a moral—imperative. For most of 1941, the States danced along a narrow, uncertain path, seeking to avoid being sucked into a "European" conflict, yet unwilling to let Hitler and the Nazis control transatlantic trade. President Franklin D. Roosevelt was particularly alive to the need to stop Hitler's aggression, and he used his inexperienced but willing navy as the instrument of increasing opposition to Germany. He gradually directed his naval forces to take more and more forceful action to prevent U-boat depredations of merchant shipping.

Eventually, of course, in spite of Hitler's determination to avoid conflict with the United States, the inevitable happened. A torpedo from a U-boat damaged the destroyer *Kearney* in October 1941, and another U-boat sank the destroyer *Reuben James* two weeks later. Still the United States stayed out of war. Only the disaster at Pearl Harbor on 7 December finally pulled the country in. And when Hitler unexpectedly supported Japan by declaring war on the United States on 11 December, the States suddenly found itself fighting on two oceans at once.[14]

In the Atlantic, the enemy campaign was directed by one man: Karl Dönitz. Admiral Dönitz controlled the U-boat fleet from 1935, when he became its head, until the end of the war. He had anticipated that in the event of another war, Britain would again resort to convoying as the best defense of merchant vessels against submarines. Therefore, combining his experiences as a U-boat veteran of World War I and his interwar experiments with torpedo-boat tactics, Dönitz evolved aggressive attack procedures. These involved the coordination of large numbers of U-boats in wolf packs in order to counteract the convoy.[15]

To assemble his wolf packs, and to guide them toward convoys, Dönitz devised a system of close, personal control over each of his U-boats, to which he gave total and unwavering commitment. "We were able, by experiment," he wrote, "to reach the positive conclusion that I could myself quite easily direct the *whole* tactical operation against a convoy from my headquarters ashore."[16] This control was to be exercised by radio, hence the argument that radio use was the key to the Battle of the Atlantic.

It is now well understood that the so-called Ultra information obtained by decryption of Dönitz's very frequent radio transmissions was the Allies' most significant source of U-boat-location information. The successful exploitation of those transmissions was primarily undertaken at Bletchley Park in England, the wartime codebreaking headquarters for decrypting intercepted German messages. The resulting Ultra information, with its origin carefully concealed, was disseminated to British and later to American operational forces as necessary.[17]

Direction-finding information, which also derived from exploitation of the enemy's use of radio, could pinpoint a U-boat's location. This made it quite feasible for convoys to avoid the U-boats, or for escorts to attack them, or both. And as submarines could not transmit while submerged, they had to surface to send Dönitz his required information regarding location and condition. This further deprived the U-boats of their invisibility, which was an increasingly serious liability as Allied air power grew in numbers and finally extended its reach to cover the entire Atlantic.[18]

Confident in the technical virtuosity of Enigma, the German electrical cipher machine, Dönitz clung to his belief in the integrity of his radio codes. His attitude toward direction finding was similar. Even before the war the Germans had known about HF/DF, which the British inventor of radar later called "our non-secret weapon."[19] But during World War I only long-wave, low-frequency radio transmissions had been detected effectively. By the late 1930s, Dönitz had come to believe that his use of the new technology of high-frequency shortwave radio would make it essentially impossible for direction finders to locate his U-boats. Consequently, even when he had considerable evidence to the contrary, Dönitz failed to take HF/DF seriously enough.[20]

In fact, U-boat Command (Befehlshaber der U-boote, or BdU) had been much more concerned with another antisubmarine technology that also had its origins in World War I. This was sonar, an acronym for sound navigation and ranging, which the British called asdic (Anti-submarine Detection Investigation Committee). Sonar was designed specifically to locate a submerged U-boat by sound waves.[21] To avoid detection by sonar, Dönitz devised surface-attack tactics for his submarines. There were few British and Canadian naval vessels available for escort duty in the early war years, and they had huge perimeters to protect; an average convoy of forty-five ships would normally cover five square miles of sea. This made Dönitz's surface tactics quite practicable.[22]

The U-boat's inconspicuous silhouette seemed perfectly designed for these maneuvers, and the element of surprise was provided by attacking at night. These night surface attacks by massed submarines became the hallmark of Dönitz's U-boat war, and for a time the tactics met with considerable success. But characteristic of the Battle of the Atlantic was the continuous spiral of countermeasures (new technologies and devices, weaponry, vessels, tactics, and intelligence), which meant that each advance on one side was almost invariably subjected to a countermove on the other.

One Allied response to U-boats attacking on the surface was to adapt the relatively new radar (radio detection and ranging) technology for use in antisubmarine warfare (ASW). Radar could detect surfaced U-boats at night as well as in poor weather. Although some progress on this technology was made in Britain between the wars, the real development of naval radar was encouraged by the need to counter the U-boats' new surface-attack tactics. Eventually radar was used to considerable effect in the antisubmarine war, though the reliability of early devices was limited.[23]

Electronics today is so sophisticated and complex a field that it is easy to forget how recent this development has been. Avery Richardson, one of a group of American engineers who worked on high-frequency direction finders during the war, has pointed out that when he was born in 1901, "there was no such thing as radio."[24] In 1939, despite rapid advances, radio electronics was a young, primitive, and often mysterious art.

During the course of World War II, there was a tremendous acceleration in the pace of development in electronics. But this was, after all, the first war heavily reliant on modern electronic devices and many people at the time were slow to recognize how essential they had become. In the midst of war, especially, the potential value of specific scientific contributions is hard to assess.

At the same time, Dönitz's blindness to the possibility of critical improvements in HF/DF was to cost him dearly. Because of the development of highly specialized and refined HF/DF devices beyond the scope of anything imagined by Dönitz and his staff, direction finders became at least as important to the ultimate defeat of the U-boats as all the other better-known electrically and electronically derived innovations. Only Ultra may have played a more important role.

Unlike HF/DF, sonar, and radar, Ultra is not an electronic device. Unlike shipborne HF/DF, sonar, and radar, Ultra never had to be made seaworthy; it never had to contend with the elements physically. Ultra was information, the end product of a highly complex and quirky process. Ultra information was produced by linguists, chess players, mathematicians, even medieval historians, rather than by the scientists and engineers who produced HF/DF, sonar, and radar.[25] It was most effective at the strategic and operational levels, and for convoy defense.

This was also true of information from land-based direction finding. But it was shipborne HF/DF that made a real tactical impact on the Atlantic campaign, and that gave escorts a considerable offensive potential. The objective of HF/DF development was to make the device as easy to read and automatic as possible, because HF/DF returns were interpreted and used immediately.[26] Like somebody giving away their location in the dark by coughing, the U-boats revealed themselves to HF/DF.

It is important to recognize the great efficacy of Ultra for the Allies in their struggle against German submarines. The successes that have been credited in large part to Ultra are a good standard against which to measure the significance of the contribution made by HF/DF to the

Atlantic victory. But no further comparison will be attempted here. Ultra was a system of intelligence while seaborne HF/DF, sonar, and radar were electronic devices used by convoy escorts operating under stress in the presence of the enemy.[27]

By early 1942, when a largely unprepared United States had just entered the war, the Royal Navy was beginning to build up a significant antisubmarine fleet of escort vessels and was developing antisubmarine doctrine and tactics. It was time for offensive action against the U-boats, and British shipborne HF/DF was developed and improved to assist in that effort.[28]

By spring HF/DF was being tested in action to some effect, and a course of training, operational procedure, and doctrine was in the process of being established. In view of the apparent British lead in this field, it is somewhat surprising to find that a separate and different American Huff Duff apparatus was developed for the U.S. Navy in preference to adopting a British device. While willing to profit from British operational experience in this field, the U.S. Navy acquired its own equipment, based on imported French technology.

Tracing the development of American naval HF/DF illustrates the complex way in which a necessary technology may be identified and produced in wartime. To be sure, HF/DF was only one small part of a formidable expansion. In 1940 the U.S. Navy Department's Bureau of Ships (BuShips) spent a total of $6 million on all phases of development, procurement, installation, and maintenance of electronic equipment on ships and ashore. By 1945 the same services cost $1 billion, both because of the proliferation of equipment types and because of a much greater number of ships.[29]

Who directed this remarkable growth? Who was responsible for deciding what electronic devices were needed by the U.S. Navy and, once the decision was made, how were those devices developed, produced, installed, used, and maintained? The strange saga of the high-frequency direction finders used by the U.S. Navy in the Battle of the Atlantic yields some surprising answers to these questions. And the widely acknowledged importance of shipborne HF/DF to the outcome of the battle makes it appropriate for a case study of the development of U.S. naval technology in wartime.

In the year before the United States was actually at war with Germany, the U.S. Navy had cooperated in the defense of trade convoys in the western Atlantic by establishing a chain of coastal high-frequency direction-finder stations and exchanging information on U-boat positions with the Royal Navy and the Royal Canadian Navy (RCN). American HF/DF research had been considerably behind that of Britain but, just as the need for this extensive shore-based HF/DF network in the United States was recognized, the design for an appropriate device appeared.[30]

In the first week of January 1941, almost literally "out of the blue," Sosthenes Behn, the eccentric and highly energetic president of International Telephone and Telegraph (ITT), arranged a conference between senior U.S. Army and Navy officers and two French engineers. These were Dr. Maurice Deloraine, head of the Paris-based Laboratoire Central de Télécommunications (ITT's French subsidiary), and Dr. Henri Busignies, Deloraine's expert on direction finding.[31]

Henri Busignies was the creator of a unique device: his instantaneous, direct-reading high-frequency direction finder. What was significant about Busignies's direction finder was the speed with which it produced a usable bearing. In effect, it worked instantaneously, by means of a cathode-ray screen. Impressed, and convinced that this device was far in advance of anything then being considered in the United States, the officers quickly decided to deploy the Busignies direction finder for use in navy shore stations. They gave ITT and its affiliates contracts to produce the device with all possible speed. Soon Busignies's HF/DF model was installed in navy receiving stations up and down the East Coast, where they began tracking German U-boats in the Atlantic.[32]

Just as the technology of naval electronics was still in its early stages in the States when Busignies arrived at the end of 1940, so too were the mechanisms for integrating the fruits of science with possible wartime needs. There was no clear relationship between civilian scientific and technological establishments, government administrative structures, and practical military requirements. In response to the European emergency there was, indeed, a proliferation of both military and civilian "control" agencies. But rivalry among these organizations and their overlapping responsibilities seemed merely to add to the confusion they were designed to alleviate.

Thus, as war approached, there had been no efficient way for the U.S. Navy to communicate its urgent electronic needs to any person or group of persons with the know-how to fill those needs. There was also no clear connection between the navy and the scientists and a vital third group: those who could actually manufacture the necessary technologies. This is amply illustrated by what happened to direction finding in the United States.

The real strength of HF/DF could not be demonstrated until a shipborne device was developed that could be used in action both offensively to hunt U-boats and defensively to protect convoys. Busignies had actually created such a device and had it ready before the U.S. Navy knew that such a thing was practicable and before it had developed the doctrine or indeed the will to use it.

The first months after Pearl Harbor were so catastrophic for the U.S. Navy that it is hardly surprising if little thought was given to the development of new offensive technologies. Adm. Ernest J. King, the new head of the navy, was primarily concerned with the war in the Pacific, although the U.S. Navy was already committed to defeating "Germany first" before turning to Japan. Believing he did not have enough escorts available to properly protect convoys, King failed to institute convoying at the outbreak of war.

Instead, the few escorts available for the protection of the eastern sea frontier were sent out on offensive anti-U-boat patrols of the sort the British believed they had already proved were futile. As a result, the U-boats Dönitz had immediately sent across the Atlantic—Operation Paukenschlag (Roll of Drums)—caused immense slaughter among the merchantmen sailing independently and unprotected along the American East Coast.[33]

About 260 ships were sunk in American waters between January and April 1942. In May King finally agreed to the institution of convoying between Florida and Maine. One of the most cogent arguments persuading him to this course was advanced by Gen. George Marshall, chief of staff of the U.S. Army. Marshall was concerned that the loss of transports caused by the U-boat onslaught would cripple his ability to ship American forces overseas.[34]

Once convoying began, the number of sinkings in the area dropped immediately as Dönitz sent his boats off for easier pickings in the Caribbean and the Gulf of Mexico. At this point King became a total convert to defensive escort of convoys and discouraged subsequent ideas for offensive action. Quite likely this attitude helped to delay the development of shipborne HF/DF for another year or so, because the device could be seen as largely offensive, unnecessary for the protection of convoys. King had to be reconverted to favor aggressive action in the Atlantic in March 1943, by which time British offensive successes were proving irrefutable.[35]

The development of shipborne HF/DF in the United States was also retarded by the unfounded paranoia of Naval Intelligence with regard to the loyalty of the French engineers responsible for the device. Even Behn "was looked upon as a European," and with the heightened sensitivities of wartime the international connections of ITT were regarded as "suspect" in some quarters.[36] Exciting but discredited claims have even been made that there was a "very special relationship between ITT and the Third Reich."[37]

Undoubtedly, administrative confusion also played a part in delaying the U.S. Navy's use of shipborne HF/DF. It was not until the tide began to turn in 1943 that some coherent organization based on analysis and advance planning began to have a substantial effect on the American war effort in the Atlantic. That spring Admiral King was pushed to create the Tenth Fleet, which directed and coordinated the struggle against the U-boats. It was only then that the U.S. Navy was well-enough organized to launch a successful, offensive antisubmarine campaign.

Fortunately, shipborne HF/DF, which the British were already proving could be an effective tactical U-boat locator, was already in the works.[38] In the spring of 1942, the U.S. Navy had authorized ITT to develop Busignies's shipborne HF/DF for use at sea. A full year before King's formal agreement to endorse offensive ASW in the Atlantic, his navy had given the go-ahead to a device that was significantly to enhance its offensive potential.[39]

In Britain, by 1942 chronic shortages and worn-out equipment inhibited industrial production of all sorts. The energetic spirit of American industry, however, fueled by vast resources, soon converted new ideas into the weapons of war.[40] Allowing for initial dislocation and confusion, the U.S. Navy eventually did a creditable job of using the country's rich scientific and industrial potential to best advantage. By comparison, Admiral Dönitz's failure to counteract Huff Duff points to his larger failure to make the best possible use of German scientific and technological genius. The Allies undoubtedly won what Winston Churchill called the Wizard War, and Huff Duff is one of the reasons why.[41]

The Allied navies needed all the technological help they could get against the formidable and very professional enemy they faced in the Atlantic.

CHAPTER 2

The Target:
Dönitz's Deadly Gray Wolves

O ne man, more than any other, shaped the Battle of the Atlantic and made it what it was. That man was German Adm. Karl Dönitz, the Lion.

Dönitz had been a submariner in the Great War and had survived the drastic military and naval pruning imposed by the Treaty of Versailles. In the interwar years he had risen steadily in the German naval hierarchy, gaining important additional experience in surface ships and staff posts. In 1935 the forty-three-year-old Dönitz was appointed officer commanding U-boats and promoted to captain.

He was still in that relatively low position in the German naval chain of command in September 1939. Then war increased the importance of the U-boat effort and brought him steady promotions. Gradually Dönitz came to direct U-boat development and doctrine in minute detail, finally exercising a degree of control unmatched by any senior naval officer in any other country and indelibly stamping the U-boat forces with the mark of his own personality.[1]

To be sure, there were other naval commanders in World War II who, like Dönitz, directed their widely scattered forces from ashore. Admirals Nimitz, Halsey and Yamamoto clearly did so. But Dönitz, in addition, was involved in all major production decisions about types of vessels built. He also had some say in the numbers produced, though in the early days he invariably had to make do with fewer submarines than he had requested. Dönitz decided on how his U-boats would be armed and he defined the tactics they would use, changing this as needed in response to changing situations.

It is true that at various times Hitler interfered directly in the strategic placement of the U-boats. In November 1941 he almost brought Atlantic operations to a standstill by insisting on diverting U-boats to the Mediterranean. In 1942 he prevented an all-out campaign against the United States East Coast because of his fear of an impending Allied attack on Norway. Dönitz opposed each of these moves in vain.[2]

In general, however, Dönitz decided where and how his U-boats would operate. He was often involved, particularly before January 1943 when he took over command of the entire German navy, in directing specific U-boat dispositions in real time, while also directing the grand strategy intended to strangle Britain's economy.[3] In fact, it is impossible to fit Dönitz's areas of responsibility into any of the conventional military categories of tactician, strategist, or grand strategist. He was all of these at once, and more. In a very real way, and in spite of Hitler's occasional frustrating interventions, the German U-boat offensive was Karl Dönitz's war.

Though outwardly the typical aloof officer in the Prussian tradition, Dönitz also had another side to his character that enabled him to relate closely to his men. His concern for them was legendary, helping to mold his almost all-volunteer corps into a force that continued to fight long after the battle on the high seas was lost. Thus the often cold, frequently monotonous, and always dangerous life of a U-boat man was generally sustained by exceptionally high morale.[4]

Because his men were scattered across the ocean in their isolated iron "coffins," Dönitz used the radio to fulfill the functions equivalent to those of a conscientious battlefield commander. That was his only way to make the rounds of his "lines"—to pump up the valor of his men, to keep some captains from shirking danger, and to rein in the recklessness of others. Dönitz was not, by temperament, the type of "château general" so infamous in World War I. Referring to commanders at the western front who set up their headquarters in remote châteaus, military historians have long noted their ineffectiveness. "Distance is . . . a negative dimension," writes John Keegan. "The man who insists on it becomes a recluse, and the reclusive commander achieves nothing. Distance must be penetrable by access either inward, outward or both."[5] In spite of the development of field telephones during World War I, those château generals failed to use available technology to bridge distance.

Dönitz would not make the same mistake. When his U-boats were at sea, radio was the only way for him to penetrate the distance separating himself and his men. Radio, he thought, would enable him to overcome the disadvantages of management from a distant command center. And indeed it did. There is no doubt that as an operational commander Karl Dönitz was one of the most effective in the war. He was also one of the most respected, even revered, by his men.[6]

Still, it is possible that his emphasis on training, on morale, on molding an elite corps, blinded Dönitz to the technological dangers of detection in an electronic age. It seems clear in hindsight that he relied too much on élan and tactical proficiency, too little on ensuring that his men would always have the technical tools to keep pace with enemy advances.[7] Because of his personal style of leadership, no one challenged Dönitz's system of command and control. This was understandable at the outset of the war when the submarine was far superior to the escorts fighting it and had things pretty much its own way.

But while the concentration of power in Dönitz's hands may have been a source of his great strength as an operational commander, that same concentration was probably a major cause of his weakness in the area of technology. Such a weakness might have been avoided by a sufficiently large and diverse staff to represent a broad array of interests and experiences. But the U-boat-command operations staff was extremely small, generally numbering about twenty. Comdr. (later Rear Adm.) Eberhardt Godt had joined Dönitz in 1938 as his chief of staff, and he remained with him throughout the war, finally serving as director of the operations branch. This partnership has been called "an excellent combination, Dönitz providing the fire and drive and inspiration, Godt the calm efficiency of the ideal staff officer." Nonetheless, a crucial weakness was the absence of "a strong, critical, analytical brain."[8]

In October 1941, Capt. Günter Hessler (Dönitz's son-in-law) became Dönitz's first operations officer, and he too held his position until the end of the war. Not only Hessler, but most of his colleagues as well, had joined the BdU staff after spending the early part of the war in command of a U-boat. In fact, Dönitz insisted that his operations staff should have had practical experience of U-boat operations in wartime, preferably that they should have taken part in attacks on convoys.[9]

Dönitz was surely correct in thinking that there would be such an identity of interest between these men of recent practical experience on his staff and the commanders at sea that he could maintain the closest possible direction of his forces afloat. But the very homogeneity of the staff, who were all enthusiastic products of Dönitz's training, as well as its small size, led to intense focus on the day-to-day, at times hour-by-hour, direction of the U-boats.

As a result the staff was overburdened with operational responsibilities, at which they excelled, but there was a corresponding lack of emphasis on reasoned analysis of enemy activities. When problems occurred, like the successful evasive routing of Allied convoys in 1941, they were addressed with ad hoc solutions such as sudden changes in tactics or shifts in area of operations. Surprisingly little sustained attention was given to determining the underlying causes of the problems. This might not have been so crucial had there been some other authority responsible for the accumulation and analysis of data and the identification of long-term trends, but there was not.[10]

To be sure, at times of crisis Dönitz and his staff considered all possible internal sources of weakness, and probing investigations were set on foot. Dönitz also had access to the advice of experts in various fields, including science and technology. But the problem seems to have been that there was no specific department or group responsible for an ongoing analysis of the Atlantic campaign to provide a steady source of informed reports to the commander, U-boats. Dönitz never had the benefit of the kind of operations research that was so significant to the later Allied campaign in the Atlantic.[11]

This was a problem that characterized the entire Nazi war effort and that has been blamed on "Hitler's distrust of the 'expert' and his propensity for the irrational."[12] For whatever reason, while the Allies employed people of high intellect to conduct extensive operations research (the British staff included two Nobel prizewinners), Nazi Germany failed to make comparable use of its own vast pool of top-notch scientists. This lack of good technical advice was to prove an obvious and dangerous shortcoming when it was time for Dönitz to assess the risk to his U-boats from radio direction finding.[13]

Throughout the war, and whatever his overall changing location and responsibilities, Dönitz maintained control over his U-boats through his uniquely small, tight-knit staff. After the fall of France in the summer of 1940, he established his headquarters at Lorient on the French coast of the Bay of Biscay. In 1942, after a successful British commando raid on the nearby port of Saint-Nazaire, Hitler forced Dönitz to move with his headquarters to Paris for safety. But this did not affect his close tactical control of the U-boats as he simply moved the whole operations room with him to its new home on the Avenue Maréchal Manoury.

Then in January 1943, when Dönitz replaced Grand Admiral Raeder as the commander in chief of the German navy (and became grand admiral himself), the submarine campaign was largely unaffected because he still retained direct command of the boats. The U-boat operations room functioned just as well in its latest home at the Hotel Steinstrasse in Berlin as it had done in Paris, and with essentially the same staff.[14]

In January 1939, Karl Dönitz had published a book called *Die U-Bootwaffe* (The U-Boat Force), in which he discussed his philosophy of submarine warfare. He predicted that the task of the U-boat in any new war against Britain would be the same as it had been in the previous war: a *guerre de course* to cut supply lines with the rest of the world. Britain would not be able to sustain a war without supply from overseas and would be forced to surrender.[15]

The submarines Dönitz planned to use in this new war were little changed from those of the previous war. They were really only submersibles, with a very limited capacity to remain submerged and with underwater movement handicapped by slow speeds. The U-boats needed a high surface speed to get to their hunting grounds along trade routes, and also to catch convoys once sighted. Their underwater capabilities had been used only at the point of attack in daylight and to hide from a pursuing escort. These requirements were met by the type VII U-boat, which became Dönitz's prime weapon in the Battle of the Atlantic. Eventually he had 628 of these boats built, more than has ever been constructed of any submarine class.[16]

The type VII was a relatively small submarine compared with the standard U.S. Navy model, and it had limited fuel-carrying capacity; therefore a limited range. The model VIIC, with additional fuel tanks, was the most effective of the type. The model VIIC had a displacement of around 750 tons, was 220 feet in length, and held a crew of 44. It carried 14 torpedoes and usually one 3½–inch, one 37mm and two 20mm surface guns. As did all submarines at the time, the type VII operated underwater on electric battery-powered motors that did not require air to function. The electric motors gave the type VII a top underwater speed of 7½ knots but an average underwater speed of only 2 to 4 knots. Even at 2 knots, however, the submarine could only remain submerged for

U-66 at Lorient, 23 March 1943.
Courtesy of Karl Degener-Böning

65 hours before it had to surface to recharge its batteries. On the surface the U-boat was propelled by diesel engines, which allowed it to sustain a speed of around 18 knots.[17] As Karl Degener-Böning, wartime radio operator on the U-66, characterizes the result of these specifications, "Generally we cruised on the surface and dove only when it was necessary."[18]

Only one other major prewar submarine design was authorized by Dönitz, the larger type IX, of which 210 were ultimately built. The type IXC displaced 1,120 tons, was 244 feet long, and had a crew of 50. It could reach a submerged speed of 7½ knots and a surface speed of 18½ knots. The type IXC U-boat was armed with between 19 and 22 torpedoes, plus one 37mm and two 20mm surface guns. But though the type IX had a longer range and was better suited to rough Atlantic conditions, it was more expensive and took longer to build, so Dönitz preferred to concentrate on producing greater numbers of the smaller, cheaper type VIIs, which continued to bear the brunt of the fighting in the Atlantic.[19]

On average, the type VII U-boat could only stay at sea for a little over a month before it needed to head home for refueling. This was to prove a problem in coordinating group tactics. One solution adopted after the war had begun, particularly for the short range and endurance of the type VIIs, was to refuel and ultimately also to resupply them at sea with everything from diesel fuel to torpedoes to fresh bread. Initially this resupply was accomplished by arranging rendezvous with well-stocked surface vessels; later by supply submarines referred to as milch cows.

While well aware of the limitations imposed by the use of existing submarine types, before the war Dönitz had to compete for resources for U-boat construction not only with the German army and air force, but also with Admiral Raeder's naval plan (the so-called Z plan), which included a large surface fleet. So his decision to stay with proven designs must have been affected by the many other insistent demands he knew were being made on Germany's industrial capacity, and by a practical assessment of how he could get the largest force as quickly and cheaply as possible. The Z plan allowed for a total fleet of 233 U-boats to be constructed by 1948, while by the spring of 1939 Dönitz had become convinced that he would need to have 300 operational U-boats on the outbreak of war to effectively block British maritime supply.[20]

While his U-boats were being built in the years before the war, Dönitz was perfecting his new tactics and was training his slowly growing fleet in their use. Dönitz planned to bring his U-boats together for mass attacks on British convoys. These mass attacks would both overwhelm the convoy escorts and sink as many of the merchant ships as possible. Dönitz's goal, according to what is referred to as the integral tonnage theory, was to secure the maximum enemy tonnage sunk per submarine per day at sea. He argued that it was the destruction not of the cargo itself but of the carrying capacity that would bring Britain to its knees. To Dönitz it did not matter what ships were sunk, so long as the tonnage destroyed was sufficient to keep ahead of Allied rebuilding.[21]

There is a certain logic to this view, or at least there was early in the war. If lost, new cargoes could be obtained, but it took long months and a vast industrial base to replace shipping. Dönitz reasoned that he could maximize the effectiveness of his limited numbers of U-boats if he concentrated them against the most vulnerable targets, without regard to what was being carried. In this way he intended to sink ships faster than they could be

Fig. 2.1. Average monthly shipping losses versus ship construction by Allied and neutral countries.
Tidman, **The Operations Evaluation Group,** *19*

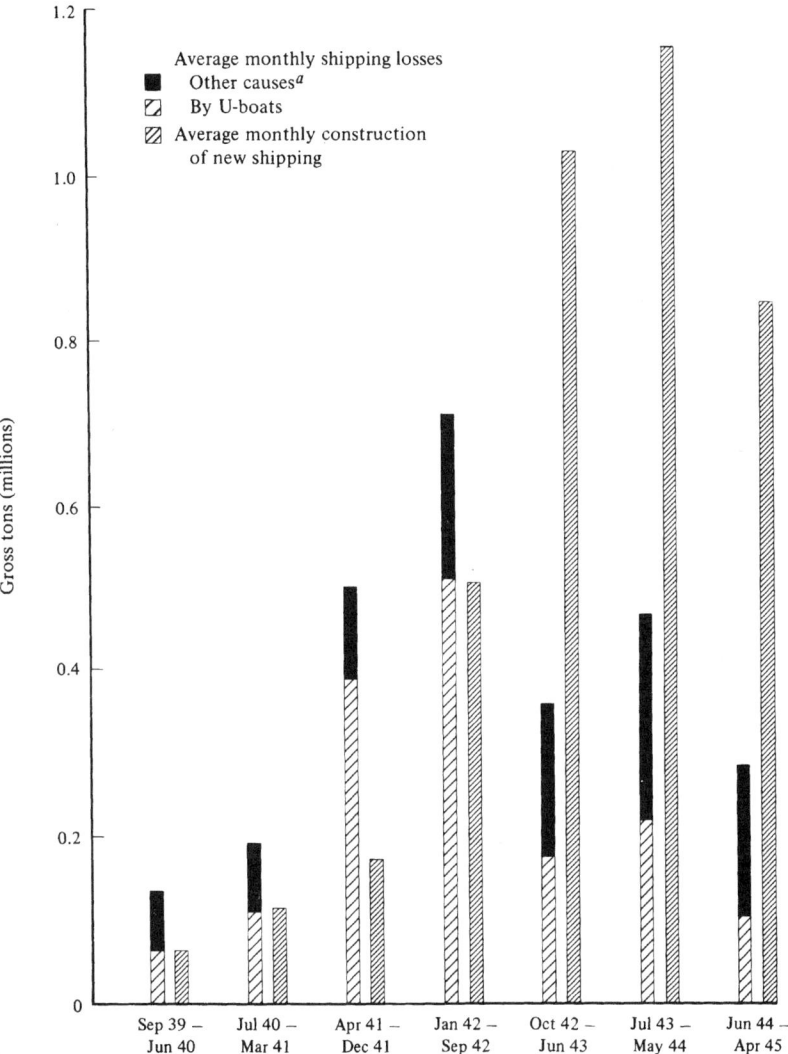

Gross tons (millions)

Average monthly shipping losses
■ Other causes[a]
▨ By U-boats
▨ Average monthly construction
  of new shipping

1.2
1.0
0.8
0.6
0.4
0.2
0

Sep 39 –   Jul 40 –   Apr 41 –   Jan 42 –   Oct 42 –   Jul 43 –   Jun 44 –
Jun 40   Mar 41   Dec 41   Sep 42   Jun 43   May 44   Apr 45

[a]By other enemy action (aircraft, ships, mines, etc.)
and natural marine casualty.

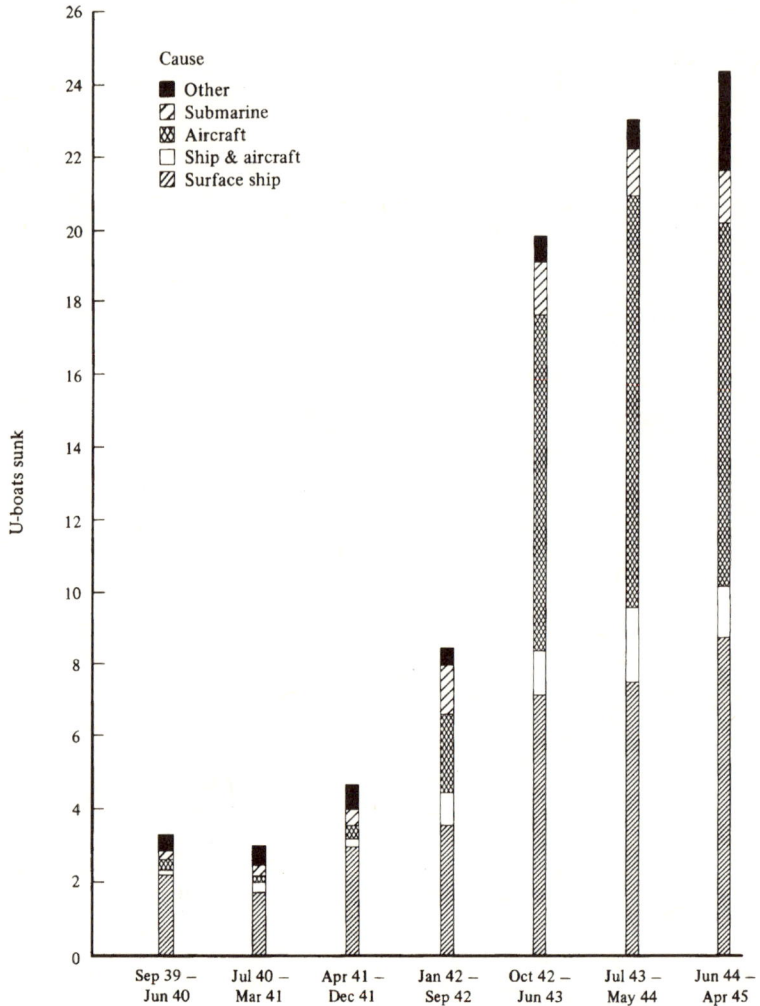

replaced, while building up his U-boat force by prudent deployment and by replacing the inevitable losses with his own new construction.[22]

Dönitz turned away from the World War I practice of using submarines as independent commerce raiders. The late adoption of protected convoys in World War I had demonstrated that lone U-boats could be neutralized when escorts ganged up on them. Instead, Dönitz planned to deploy U-boats in groups that became known as wolf packs. These groups would all be controlled by radio directly from U-boat headquarters.[23]

The vastly improved radios of the 1930s, especially those using high-frequency transmissions, promised to make possible entirely new strategic and even tactical plans. So Dönitz devised an unprecedented system of command and control that relied completely on frequent radio transmissions between U-boats and headquarters. He realized he needed close radio contact to maintain tactical control of his U-boats, and above all to direct them effectively onto convoys in coordinated mass attacks.[24]

Once a wolf pack had been assembled and was in contact with a convoy, each boat would attack independently, usually at night and on the surface. These surface attacks would hamper British asdic (sonar), which was considered the most serious offensive threat, and the U-boat's low profile was counted on to make it almost invisible at night. Being on top of the water already, the U-boat could use its surface speed to dash away from the convoy and escorts if it was spotted. Continuing ahead on the surface, away from the initial attack point, the U-boat would try to position itself where it could submerge, if necessary, and wait for the convoy to catch up once again so that it could renew the assault. Night surface attack by a coordinated group of U-boats acting like ocean-going motor torpedo boats was the essence of wolf-pack operations. While such tactics would eventually prove highly successful, they were not without their own special danger of detection.

Dönitz was well aware of how his system of radio control could reveal the location of his U-boats to their enemies. In fact, he even notes in his *Memoirs* that some top officers in the Naval High Command opposed his prewar plans to use group tactics precisely because they necessitated too much dangerous use of radio. Dönitz explains that he, "on the other hand, held the view that the mainte-

Fig. 2.2. Average monthly U-boat sinkings by cause of sinking.
Tidman, **The Operations Evaluation Group,** *25*

nance or breaking of wireless silence should be regarded purely as a matter of expediency and a means to an end and that the disadvantages inherent in the breaking of radio silence should be accepted when the use of our wireless would enable us to mass our U-boats and thus achieve greater success."[25]

U-boat Command generally sent out twenty to thirty messages a day, repeating each one two, six, twelve, twenty-four, and sometimes forty-eight hours later. The very number of transmissions required for group operations exposed Dönitz's U-boats to three kinds of danger. The first was from analysis of radio traffic, which is interpreting the origin, volume, and types of messages for clues as to their meaning. The second was from decryption of cipher messages, and the third was from radio direction finding.

Of the three forms of danger, Dönitz was most concerned about the security of his naval ciphers. Fear of betrayal or infiltration by spies encouraged him to keep his staff small and to order periodic investigations of all security systems, but not to reduce transmissions. The result of each investigation was to assure Dönitz that there was no evidence that naval ciphers had been broken, although in fact, of course, they had.[26]

The sheer number of the transmissions also increased the exposure to direction finding. While he was well aware of the network of British radio interception and direction-finding shore stations built before the war, Dönitz was not unduly afraid of them. He was, after all, repeatedly assured by his staff that it was impossible to obtain directional bearings on the brief shortwave (high-frequency) transmissions that the U-boats would be making. Of course, his advisers were wrong. But in 1958 Dönitz still wrote with misplaced confidence that "the introduction by the U-boats of short code signals, so short that the enemy D-F [direction finder] could locate the transmitter's position only very approximately, if indeed at all, proved to be of great value."[27]

In hindsight it is surprising that Dönitz did not take such multiple vulnerability seriously enough to order a substantial reduction in the number of transmissions until late in the war. At times caution was recommended, but still the demand for information usually overrode qualms about detection.

After the war the British Admiralty had Dönitz's son-in-law, Günter Hessler, write an account of the Atlantic campaign based on official German naval records. Hessler noted that during the invasion of Norway in April 1940, "The U-boat Command demanded an explanation by radio. There was no objection to transmitting, as the risk of being 'D/F'd' [located by direction finding] could be accepted, and the enemy's knowledge of the presence of U-boats would have a deterrent effect on him."[28] By forcing the U-boats to use their radios, thus revealing their presence and even their precise location, Dönitz not only deprived them of their stealth, he eventually made them highly vulnerable to counterattack.

None of this was obvious in 1939, however. And Dönitz's brilliant innovations in submarine tactics created a superbly effective force that, by 1943, had badly shaken the Allies. Dönitz's early success is all the more surprising because, relying on Hitler's promises, he was not expecting a war before 1944.

In the first phase of the war, Dönitz could only keep about ten U-boats in their operational areas at any one time, and this was too few for pack attacks. So the Atlantic campaign began with the U-boats patrolling individually to intercept Allied merchant ships and making traditional, independent, periscope-depth attacks in daytime. There was little radio communication, as there were no groups to be coordinated.

At this early stage asdic was almost the only source of U-boat-location information available to the British. The Government Code and Cipher School at Bletchley Park had made a good start at unscrambling some of the German ciphers encrypted on their Enigma cipher machine. In early 1940 combined British, French, and Polish efforts had led to breaking the Luftwaffe general-purpose key called Red. But the German naval version, Schlüssel M, used more rotors and could not be broken at this time, especially given the paucity of transmissions with which to work. This silence from the Kriegsmarine also greatly reduced the significance of the already operational British shore-based direction-finder network, which could produce no results if there were no transmissions.[29]

Thus, until about June 1940, because both sides lacked the organization and the vessels to mount group operations, there was relatively limited use of radio communication and therefore relatively little availability of radio intelligence. This situation changed rather dramatically in July 1940, by which time escorted convoying was well established and was countered by the opening of the first real U-boat onslaught. In the summer of 1940 the Germans began to use the radio extensively to direct both U-boats and surface ships onto the convoys. On the Allied side, efforts to avoid wolf-pack attacks on convoys led to rerouting, which greatly increased the volume of their own radio traffic.[30]

The pattern of reciprocal moves and countermoves resulting from the first extensive use of radio brought into play the various tools each side developed to turn the enemy's transmissions against him. Now radio-communications intelligence began to have an influence on the evolution of the fighting in the Atlantic. This influence was still weak, though; the only radio intelligence available to the Allies was shore-based D/F and analysis of radio traffic. But as Dönitz predicted, the D/F'ing was not accurate enough at a range of over two to three hundred miles to ensure routing convoys clear of a gathering pack. Nor were any significant decryptions available yet.[31]

On the other hand, the Germans had already broken into British naval codes before the war began, and the U-boats were soon relying on the B-Dienst's interception and decryption of British radio messages to overcome the problem of finding targets in the vast ocean. By April 1940 the B-Dienst (the cryptanalysis section of German naval intelligence) was able to read a third to half of all intercepted British naval cipher messages, many of them convoy routing instructions. This information was extremely valuable to BdU because the convoys were operating in a three-million-square-mile region. In spite of their relatively larger size, convoys were harder to find than great numbers of ships sailing independently.[32] Meanwhile, British land-based air opposition to German attacks on merchant shipping, though weak at the beginning of the war, had gradually intensified, edging the U-boats farther out into the Atlantic. The conquest of Norway and then the fall of France in the summer of 1940 relieved this strain on the U-boats by giving them access to Norwegian and French Biscay ports. The Norwegian proximity to the North Atlantic and, more important, the westerly position of the Bay of Biscay, extended the range of the U-boats, enabling them to reach patrol areas well away from the land-based air cover that was now making operations close to Britain too dangerous. Dönitz constructed U-boat bases with impregnable concrete pens all along the Biscay shore, at Brest, La Pallice, Bordeaux, and Saint-Nazaire, as well as at Lorient.[33]

The conquest of France solved the problem of extending the range of the U-boats into the Atlantic away from Allied air. But lack of cooperation from the German air force in the location of targets was a con-

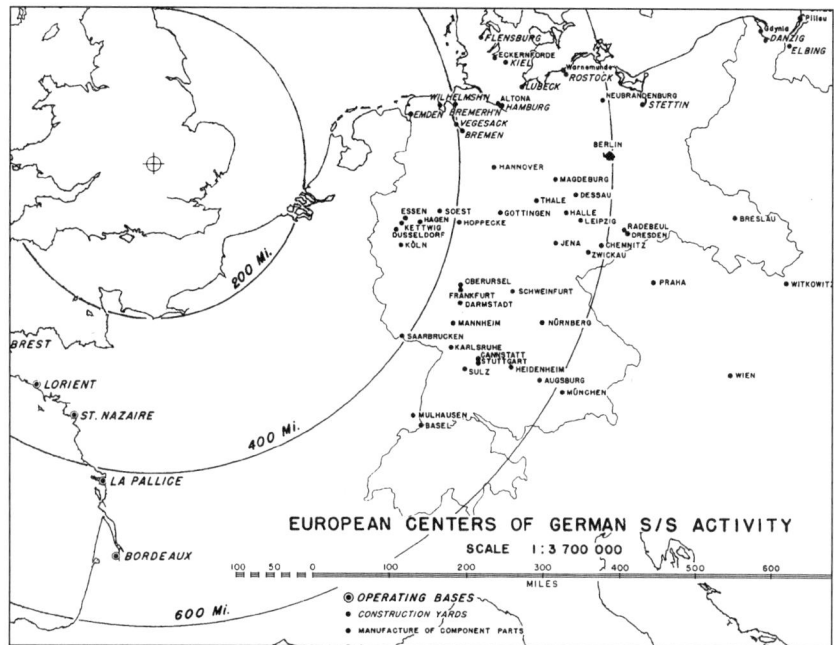

Map 2.1. European centers of
submarine-service activity.
NHC/OA, Double Zero Files, Box 56

tinuing source of aggravation for
Dönitz. Hitler increased the normal friction between the various
services and encouraged counterproductive competition by his practice of having conferences (the so-called Führer conferences) with each
of his service chiefs individually. This made it more difficult for Dönitz
to secure the cooperation of the air force.[34] Combined operations were
somewhat successful in the Arctic, as the Luftwaffe often kept numbers
of aircraft there that were suitable for anti-shipping operations. But because he had to struggle to obtain even inadequate air reconnaissance
in the Atlantic (where Göring resisted assigning aircraft to work with
him), Dönitz came to rely more than ever on B-Dienst intercepts as well
as on sighting reports from his own U-boats.[35]

Typically, a group of U-boats would leave their Biscay ports one at
a time and make their way to specific locations in the Atlantic. As soon
as they had cleared the bay, they signaled headquarters so that Dönitz
knew they were on their way. Since Dönitz could rarely send his boats
to sea with full operational orders, once they were in the predetermined

operational area they received instructions from headquarters by radio, usually to form into a scouting or patrol line. A scouting line was a line abreast with a distance of about twenty miles between each boat. The exact position of this line was usually determined by the most recent convoy information from the B-Dienst, when this was available. Thus the line was intended to be centered on, and at right angles to, a suspected convoy route. As the number of operational U-boats grew, so Dönitz's patrol lines became longer and longer, increasing the chances of intercepting convoys.[36]

As soon as any U-boat sighted a convoy, it was required to make an initial report by high-frequency radio transmission to shore receiving stations indicating the course and the speed of the convoy. The U-boat then shadowed the convoy, making amplifying reports with any other information it could acquire, such as the number of merchantmen in the group and the number and type of escorts. This boat also acted as a homing beacon, sending out a medium-frequency (MF) signal, though only to a range of sixty to one hundred miles, to guide the rest of the patrol line onto the quarry.[37]

In the meantime, information from the shadowing U-boat or boats went to BdU, where tactical deployment of each submarine was carried out under the direct supervision of Dönitz himself. An exact repetition of the initial sighting report was rebroadcast from headquarters and was directed to the attention of all submarines in the area. The captain of each submarine then estimated whether or not his vessel was in a position to intercept the convoy in time to attack it. The U-boats reported back to headquarters with their position, and also with information on their general condition, their fuel situation, and the number of torpedoes they had left.[38]

BdU considered all of these reports when organizing an attack. Then Dönitz instructed each of the U-boats he chose for the operation as to the course and speed required for making contact with the convoy. Dispositions were designed to bring the maximum number of submarines within striking range. When they approached close enough, the U-boats could take MF/DF bearings on the homing signal from the shadowing submarine in order to check their progress toward the target. But the submarines rarely communicated with each other, and they were under strict orders (which became increasingly emphatic during the course of the war) to maintain radio silence when they started to move toward the convoy, so as not to give away their presence to Allied D/F. Once the attack was under way, however, there were very few limitations on the use of the radio. The reasoning behind the relaxation was that stealth had already been forfeited.[39]

Even though transmissions were severely restricted at certain times, in general the U-boats were characterized by their extensive radio "chatter." But high-frequency radio communication from the U-boats was always directed only to headquarters control and not to nearby comrades, which gave each boat a sense of isolation even if it was a member of a pack. Peter Cremer, who later served on Dönitz's staff and who was one of the few veteran U-boat captains to survive the war, has left an account of what this policy meant to the submariners: "Even in a pack, one boat knows next to nothing about the others unless the radio signals are analyzed or the group concentrates to within visual distance, which seldom occurs. One imagines oneself alone in a wide space, a small dot in the endless Atlantic. But U-333 [his boat] was always only one of the boats directed to a convoy by radio after the first sighting reports."[40]

So the radio was not used to maintain a sense of unit cohesion within the wolf packs. In fact, Dönitz himself has left the best record of why he required his submariners to communicate with him:

> If, as sometimes occurred, my "on-the-spot picture" seemed to be insufficiently clear, I used to ask by radio for further information, and as a rule I received an answer within half an hour. Whenever I was called upon to make some special decision, which depended upon precise knowledge of any particular given circumstances, I used to speak personally in code language to one of the commanders whom I had previously informed by radio of the precise time at which I would call him. I would then be put completely in the picture.[41]

Although Dönitz adds that these "special procedures . . . were only used on rare and isolated occasions," it is clear that he expected, as a matter of course, to hear from his U-boats frequently, and in a prompt and timely fashion.[42]

BdU Standing War Order No. 2, of 2 September 1941, states clearly: "Every U-boat in contact with main targets is to make regular contact reports."[43] Sending specific sighting and engagement information put the U-boats in considerable danger. But Dönitz also required them to send him regular reports not only about position, heading points, damage, fuel and torpedo status, but also regarding performance of new equipment, requests to return to base, weather reports, and other information of all sorts.[44]

By contrast, the U.S. Navy's submarine force, the so-called Silent Service, was impervious to Japanese direction finding in the Pacific because each boat operated individually, maintaining strict radio silence.[45] Without doubt, the U-boats were extremely vulnerable to radio D/F because of their frequent transmissions, and this vulnerability was to prove instrumental in their defeat. The degree of Dönitz's control of radio use is demonstrated by the general instructions to his U-boats cited in his memoirs:

> In the actual operational area: Radio to be used only for the transmission of tactically important information, or when ordered to do so by U-boat Command or when the position of the transmitter is in any case already known to the enemy.
>
> En route to or from the patrol area: As above. Signals of lesser importance may be sent, but only very occasionally; in this connection care must be taken that the transmission does not compromise the area for other U-boats either already in the area or on their way to it.
>
> Technical: Frequent changes of wavelength, additional wavebands and wireless discipline to add to the enemy's difficulty with D/F.[46]

To be sure, U-boats were instructed not to signal receipt of messages in order to cut down on the number of transmissions they made and to avoid giving clues as to their location. Nevertheless, the emphasis of Dönitz's guidelines seemed to be much more on transmitting with care than on avoiding transmission. Clearly too, no danger was perceived from direction finding once the U-boat's presence was known to the enemy. There was no suspicion of the tactical use of shipboard D/F to locate and attack nearby U-boats.

Karl Degener-Böning served as radio operator with the rank of petty officer on board the U-66 from its commissioning in January 1941 until it was rammed and sunk by the U.S. destroyer escort *Buckley* on 6 May 1944. He describes his radio duties this way:

> Mostly we received and transmitted radio messages while on the surface. We were always receiving on different shortwave frequencies. Therefore we had several fixed-frequency schedules, with a lot of dif-

U-66 petty officers on USS *Buckley* (DE-51) after the clash of the two vessels, 6 May 1944. Radio operator Karl Degener-Böning is seated second from left.

U.S. Naval Historical Center

ferent frequencies for different areas assigned to all subs in that area. If we had to change the schedule we got orders from HQ. To follow our schedule we had to change the frequency every two hours. We had to take every message that came over this ordered wave. Only after decoding the message could we see whether it was of general importance, or specifically important for us. As every message had a number, we could see whether we got all of them. Every message was repeated several times on all waves. If we had been forced under the surface for a long time and had missed some numbers, we had the possibility of catching up on the special longest-wave frequencies. These we could receive below the surface at periscope depth of approximately twenty meters.

To transmit radio messages, generally we used the same frequencies that we received on at that time. So we got the transmission acknowledged at once. Another possibility was to transmit on other waves where we could be picked up from special stations that were listening all the time. The content of a message was given to me by the commander; we wireless operators had to code the message and transmit it. Very secret messages had to be coded and decoded by an officer. . . . Our

The U-66, which was destroyed by USS
Buckley on 6 May 1944.
U.S. Naval Historical Center

radio communications were by
Morse signals, and the content was
coded in groups of four letters
with the Enigma device.[47]

The radio and sound equipment that Degener-Böning remembers
having on board the U-66 consisted of:

  2 200-watt shortwave transmitters
  1 150-watt long-wave transmitter
  1 40-watt emergency transmitter
  2 shortwave tuners
  1 long-wave tuner (used also as a sound tuner, to sound
        radio signals from land stations)
  1 Atlas multi-unit hydrophone for listening
  Several radar tuners for different frequencies, to detect
        radar impulses [radar search receivers]
  One radar transmitter. We made no good use of it.[48]

Also in the radio room, though not mentioned by Degener-Böning,
would have been the Morse key bench, the Schlüssel-M (naval Enigma
cipher) machine, and a phonograph. There were two radio/sound men

in the crew (as well as two emergency backups). When the U-boat was on the surface, the two men would take turns on watch in a pattern of four- and six-hour shifts. When it was submerged both men would be at work together, one on the hydrophone and the other at the radio.

Wolfgang Hirschfeld was a U-boat radio operator from December 1940, when he joined the crew of the U-109, until the surrender of U-234, on which he was serving at the end of the war. He has made an invaluable record, in the form of a daily logbook, of all his duties on board. Above all, this confirms the frequency and the persistence of U-boat radio use.[49]

The German navy used four main types of messages, each with specific indications and order of precedence. Because they had different prefixes it was easy to distinguish between messages sent out from control stations and those transmitted from the U-boats. There were also different procedures for sending homing signals, for signaling aircraft, and for U-boat/supply boat cooperation. Standardized messages were preceded by a Greek letter, repeated. Operators transmitting in the naval cipher Triton, for example, always preceded the messages with Beta Beta. Triton (which the British referred to as Shark), was used beginning in February 1942 for signals between the Atlantic U-boats and Dönitz's headquarters.

Allied trackers came to call Beta Beta messages B-bar, because they were similar to the British B-bar Morse symbol. The U-boat radio operators also adhered strictly to the German naval frequency plan and later to a standard method for shifting frequencies. This regularity greatly assisted the Allied HF/DF technicians. Eventually a detailed syllabus was established for courses to train Allied HF/DF operators in German radio procedures.[50]

While Dönitz failed to substantially reduce the number of transmissions he required his U-boats to make, at least he did encourage keeping those transmissions as short as possible. There were two main ways of doing this: by compressing the message and by compressing the transmission itself. From early on in the war the German navy, and especially the U-boats, used a codebook, the Kurzsignalheft, to reduce the length of standardized radio messages. One letter was substituted for an extended text. The whole brief coded message was enciphered by the Enigma machine, letter by letter, and was then ready for transmission.[51] Thus the ten-word message "No enemy traffic. Strong defense.

Oberfunkmeister Wolfgang Hirschfeld, on the U-234 from January 1943 to August 1945.
Courtesy of Wolfgang Hirschfeld

Damages. Return to base. U-953" could be sent in only four letters.[52] These "short signals" were confidently expected to be too brief, even if detected, to be used to plot a line of bearing on the transmitting U-boat.

Still, during the course of the war, the fear of enemy D/F gradually increased among U-boatmen, and existing radio procedures were adjusted to deal with the apparent growing danger. Frequency changes were used more often, though both sides were aware that their main frequencies were well known and that all traffic was routinely monitored. A method was therefore sought to make transmissions the enemy would not even be aware had taken place. The answer was believed to have been found in the procedure known as Ausserhalb der Schwebungslueke, which the Allies referred to as off-frequencies.

Off-frequencies involved transmitting on either side of the normally specified frequencies, plus or minus, according to a variable and elaborate set of tables, the key to which was transmitted from only one shore station, Norddeich Radio.[53] However, the Allies soon constructed charts for HF/DF operators giving these German off-frequencies to guard, which indicates that procedural changes were not a very effective deterrent to radio location. While patterns and practices might change, the fundamental weakness remained the continuing insistence on transmission. The reason for this lay in Dönitz's fixed determination to maintain the closest tactical control of his U-boats, and he could not see doing that without constant mutual exchange of information.

Germany too had shore-based HF/DF. But because of the restricted geographic area available for the location of German HF/DF shore stations, they could not often get clear bearings even from nearby transmissions, and only poor cross-bearings at any greater distances. This compared unfavorably with the huge longitudinal spread of the Allied stations along the western Atlantic from Canada to the southern Caribbean. Thus their own limitations may have given German engineers a false sense of the weakness of the device.

Nevertheless, the B-Dienst was manning sixteen intercept stations on the outbreak of war, and later on German direction finders did occasionally get useful bearings from stations in Norway and France. This was the case in February 1943, for example, when convoy ON 166 was located from signals sent by its protecting aircraft. But such occurrences were not significant enough to wean Dönitz from reliance on radio coordination.[54]

The U-bootwaffe was certainly not the only German force insufficiently sensitive to the dangers of radio use. In the Mediterranean, Allied submarines supported by Ultra played a significant role against the Regia Marina Italiana, the Royal Italian navy. The Italian C38m, or Enigma code, carried convoy information, and the British broke the code in the fall of 1940. This might not have been fatal to the convoys, as Italian naval operational orders were usually sent by wire between Italy and Africa. The Italian air force also generally avoided operational radio transmissions. But both the German air force and the Afrika Korps regularly used the radio for such transmissions, erroneously convinced that C38m, like their own Enigma codes, was unbreakable.[55]

It has even been argued that use of the radio by German forces in the Mediterranean "helped to compromise the whole Italian war effort."[56] Perhaps a healthy dose of paranoia about radio use in general, rather than what proved to be an exaggerated confidence in Enigma codes, would have served Germany better. As late as 1944 German scientists were studying infrared, ultraviolet, heat-seeking and other techniques of ship location, but they undertook no comparable study of HF/DF.[57]

German experts also continued to hold to the belief that it was impossible to produce a shipboard HF/DF set because the size and weight of such gear would be too much for small boats like the destroyers on which it would need to be deployed.[58] They had harnessed MF waves and had MF direction finders on their U-boats. They used MF transmissions from submarines in contact with convoys to make homing signals "on which the remaining U-boats concerned will take D/F bearings and this will enable them to check their course and speed."[59] Still they remained convinced of the insuperable problems posed by the techni-cal differences between medium-frequency direction finders, which were easy to install on ships, and high-frequency direction finders, which were not.[60]

In addition to issuing instructions to his U-boats and demanding information from them, Dönitz also used the radio to give his subma-riners news from home. After the transmission of regular tactical in-formation, headquarters would often send the latest military news. There were also frequent personal broadcasts, which were good for morale. Peter Cremer records two typical examples. Chief Mechanic Heber on board the U-333 received the message: "Healthy girl arrived. Mother and child well." A similar notice to another new father an-nounced the arrival of a "small sailor with periscope."[61]

All of this made for great activity in the airwaves. But Dönitz confi-dently continued to underestimate the danger to his U-boats from their radio transmissions, secure in the notion that the sophisticated cipher machine and complicated codes for everything from geographic loca-tion to U-boat designators were the best way to avoid discovery, and thus to maximize surprise.

At the end of May 1941 there occurred another one in a series of radio-intelligence breakthroughs that punctuated the development of the campaign in the Atlantic. The British intelligence officers at Bletch-ley broke the German cipher key M-3, the naval home waters version that they codenamed Dolphin. They could now regularly, though with some delays, read orders from Dönitz to his submarines organizing the long patrol lines to intercept the convoys.[62]

Most important, the British had an effective system in place to make good use of the Ultra information provided by decryptions. Intercepted German transmissions were sent from the intercept stations to Bletch-ley Park for decoding. Then, in English translation, they went by tele-printer to the Operational Intelligence Centre (OIC) in the Admiralty in London. In the OIC the information was passed on to the Submarine Tracking Room, where it became part of the whole intelligence picture of the war against the U-boats.[63]

Ultra intelligence was especially important because it could reveal the Germans' intentions and their dispositions. Sightings of U-boats and ship sinkings also provided valuable input. Shore-based, long-distance direction finding, another main source of information, could pinpoint a U-boat's actual location from the direction of arrival of its radio transmission. All of this data was combined and evaluated and then entered on what was known as the Submarine Position Plot, which was sent out daily to operational commanders. The information from the plot could be used defensively to reroute convoys, improving the chances of the merchant ships and the "safe and timely" delivery of their precious cargoes.

Based on all sources of information, but especially on Ultra intelligence, the Submarine Tracking Room rerouted convoys around U-boat positions so successfully during this next phase of the battle from May 1941 to January 1942, that by one estimate it saved three to four hundred merchant ships, or 1½ to 2 million gross tons.[64] This savings represents between 10 and 14 percent of the total number and tonnage of merchantmen that U-boats sank in all areas over the entire six years of the war. Loss of an additional 10 to 14 percent at this stage, before the United States was officially engaged in the war, might have proved decisive. Radio intelligence, in this case Ultra information, was already playing a key role in the Atlantic, but the battle was not yet won.

It was during this critical phase that the U.S. Navy was finally drawn in. On 11 September 1941 Roosevelt announced in a broadcast that "From now on, if German or Italian vessels of war enter these waters [Iceland and similar areas under United States protection], they do so at their peril."[65]

While far from fully prepared, the U.S. Navy's support force at once began active convoy-escort operations in cooperation with the Royal Navy and the Royal Canadian Navy. On 15 September the U.S. Navy announced "that it would provide protection for ships of every flag carrying land-aid supplies between the American continent and the waters adjacent to Iceland, on which a U.S. base had been established in July 1941."[66] At the same time the United States took over operational control of the Allied convoy operations in the North Atlantic west of twenty-six degrees west.

To direct the new joint British, Canadian, and American convoy operations, the Allies introduced the inter-Allied naval cipher 3, and the B-Dienst achieved some breaks in this cipher quite quickly. But this had no effect comparable to the British breaking of the German M-3 key in May. That accomplishment had enabled the Submarine Tracking Room not only to save the convoys, but also to direct forces in a concerted attack on the tankers and supply boats for German surface raiders. Seven of the eight support ships had been sunk by 21 June, six of them directly due to Ultra information. The success of these operations effectively ended surface raiding against merchant shipping.[67]

There were two important consequences of the loss of the German supply vessels. On 9 June 1941, Dönitz had issued Standing Order 243 as the direct result of inquiries by his staff and by the B-Dienst about the risk to transmitting U-boats from British direction finding. Because the inquiries now indicated that reasonably accurate fixes could be obtained from many transmissions, the order instructed all U-boats in approach and attack areas to limit their transmissions to those that were tactically important and, significantly, to responses to questions from headquarters. Again, no restrictions were imposed on transmissions once the enemy was aware of the U-boat's presence.[68]

Meanwhile, the British were afraid that the diversion of convoys, as well as the extraordinary coup against the German supply system, would alert the enemy to the possibility that their cipher had been broken. They therefore determined that they could not afford to use Ultra except under cover of other information sources. This shielding of the origin of radio-intelligence information was built into the system that disseminated the information for operational use. It is now very difficult to sort out, for example, what information derived from prisoner interrogations, traffic analysis, or HF/DF fixes, and what was Ultra. Protection of the source of Ultra information was also adopted in the United States, which has similarly covered up its historic trail. What does seem clear is that U-boat locations stated as deriving from HF/DF fixes had sometimes really been obtained from decryptions.[69]

However the information on the location of U-boats was obtained, the business of ferrying supplies across the Atlantic was still difficult and dangerous. Though three months were to pass before the United States was officially at war with Germany, from September 1941 the U.S. Navy had become an increasingly active participant in the British and Canadian antisubmarine effort. On 17 September Grand Admiral Raeder reported to Hitler that "there is no longer any difference between British and American ships," and he asked for permission to authorize attacks on escorting forces "in any operational area at any time."[70] But Hitler was preoccupied with the Russian campaign and continued to urge "that care should be taken to avoid any incidents in the war on merchant shipping before about the middle of October."[71]

On 17 October the USS *Kearny* was torpedoed. It had been one of a division of four American destroyers sent to screen a Britain-bound convoy south of Iceland. Ben Brooks, the *Belknap's* communications officer, watched the *Kearny* enter Hvalfjordhur harbor and tie up to the *Vulcan*, a destroyer tender originally sent to Iceland to make an emergency replacement to the *Belknap's* port propeller. A few hours after decoding a lengthy casualty report from the *Kearny* identifying eleven seamen dead and twenty-four wounded, Brooks was startled to catch a radio broadcast in which he "heard Secretary of the Navy Knox tell an anxious nation that he had just heard from the ship and that there were no casualties."[72] So much for truth, the first casualty in war. Apparently neither side was willing to completely abandon the fiction of American neutrality just yet.

Hitler continued to approve directives "to lessen the possibilities of incidents with American forces," but in the murky and confused business of Atlantic warfare, further clashes were inevitable.[73] On 31 October, also south of Iceland, the old U.S. four-stacker *Reuben James* was blown up after a German torpedo hit in the forward magazine. But still there was no declaration of war.

Pearl Harbor changed everything for the Atlantic as well as the Pacific. On 11 December Hitler joined Japan by declaring war against the United States. The next day Brooks saw "a large formation of battleships, cruisers, and their destroyer screen leave Iceland for the Pacific."[74] The requirements of a two-ocean war limited the effectiveness of the U.S. Navy, especially in the Atlantic, until the great surge in ship construction could begin to make itself felt.

Ultimately, the tremendous American industrial capacity ensured the inevitable failure of Dönitz's tonnage war. But in the meantime, when the United States finally entered the war officially it made matters better, instead of worse, for the U-boats. January 1942 heralded what the U-boatmen called their Second Happy Time of almost unopposed attacks, this time against unprotected shipping in American coastal waters. The advantage swung still further to the German side when on 1 February the Kriegsmarine introduced the new M-4 Enigma machine so the U-boat cipher Triton could no longer be read. This caused a special-intelligence blackout for the Allies that lasted most of the year.

Even without decryptions, however, there were several other kinds of intelligence that could be extracted from whatever enemy transmissions were intercepted. It is possible to distinguish the Morse signature, or sending characteristics of individual radio operators, and the British had been keeping a record of those characteristics for each of the known U-boat operators. This information could be used to identify the where-abouts of specific U-boats. Sometimes it was even possible to distin-guish the electronic signature of individual transmitters, which could be used in the same way to keep track of a U-boat's location and move-ments.[75]

In spite of the Ultra blackout, therefore, the Submarine Tracking Room had some information about the number and routes of German U-boats at sea during this phase, and in fact in January they tracked them inexorably moving across the Atlantic toward their unprotected prey. The British had been passing invaluable Ultra information to the United States for some months past. But when they sent word of the approach of the first U-boat assault wave, it still took the U.S. Navy five months before it instituted the Interlocking Convoy System.[76]

As a result merchant-shipping losses were devastating, and Admiral King has usually been blamed for them. A more balanced view would also give due consideration to the general unpreparedness of the United States for war, the relative inexperience of the U.S. Navy, and the very real difficulties of prosecuting a far-flung, two-ocean campaign.[77]

The long-distance U-boat operations in American coastal waters were greatly facilitated at the end of April when the first U-tanker, U-459, went to sea. This huge 1,700-ton type XIV supply boat carried no offensive weapons but could dispense up to 600 tons of extra fuel oil to other U-boats on patrol. For this reason the U-tankers were quickly nicknamed milch cows. Refueling operations could make it possible for twelve me-dium-size boats (type VIIs) to operate in the Western Caribbean, or could extend the effective radius of five large boats to as far south as the Cape of Good Hope.[78]

The great advantage of the increased range that the milch cows pro-vided to the U-boats was tempered by the danger of interception of the transmissions required to effect a rendezvous, and by the often inevi-table concentration of several boats milling around waiting to be re-fueled. Eventually, ten large supply submarines were built of types XIV and XB. But the use of the radio was once again the Achilles heel of tac-tical coordination of U-boat operations. The inherent risk of discovery during supply operations was so great that by the end of 1943, Dönitz had lost every one of the milch cows as well as numbers of the U-boats meeting up with them.[79]

Both Ultra and shipborne HF/DF played essential but significantly different roles in the defeat of the U-tankers. Ultra information was of limited use in actual tactical operations because decryption usually took too long to provide real-time data. But the information about U-boat habits and refueling locations that the analysts culled from decryptions enabled them to guide the search for milch cows to the right general area with considerable accuracy.[80] Once in the vicinity, information from shipborne HF/DF could be used to vector aircraft and escorts directly onto the U-boats at rendezvous. Thus much of the success against milch cows was due to the effective deployment of seaborne HF/DF.

By May 1942, when convoying was finally introduced in the eastern sea frontier and air coverage was improved, the U-boats that the U-459 had refueled were moved into the Caribbean and the Gulf of Mexico. They were successful in those areas for most of that summer, until defenses were organized there too. At that point Dönitz moved his U-boats back to the North Atlantic. Here the U-boats now concentrated on the midocean Air Gap south of Greenland (sometimes called the Greenland Air Gap, or by Germans the Black Hole). This was the area beyond the range of most land-based airplanes taking off from Europe, Iceland or America. Newfoundland would not become operational as a base for the very long-range (VLR) aircraft used for ASW work until June 1943, and in spring 1942 there were fewer than two dozen VLR B-24 Liberators serviceable in Iceland and Northern Ireland.[81]

Because there was no fear of attack from the air in the Black Hole, Dönitz could assemble large wolf packs of up to thirty or forty boats within this zone and have them safely converge on convoys on the surface. These tactics, if pressed home, often overwhelmed even numerous escorts. In fact, it was found that when U-boats outnumbered surface escorts by more than a two-to-one ratio, it was impossible for the escorts alone to ward off concerted attacks.[82]

By July 1942 renewed and extensive use of radio once again became necessary for the operational and tactical control of the large U-boat groups now deployed by Dönitz. That same month the U.S. Anti-Submarine Warfare Unit issued a study of the U-boats in which it characterized their campaign as "a constantly changing one, the location and activities of each boat being controlled in mechanical fashion from their home station, possibly in occupied France."[83] On the Allied side the renewed onslaught in the North Atlantic also led to increased radio use to control, direct, and redirect endangered convoys. Once again there was a great role to be played by radio intelligence.[84]

By the end of 1942, with everything that could be learned from the B-Dienst, Dönitz was able to position his patrol lines so accurately that, increasingly, convoys headed right into the middle of them. This enabled more of the U-boats to come up for the attack on the very first night of contact.[85] On the Allied side, moreover, British cryptanalysts had not yet broken the current German naval cipher (M-4, Triton), so the long blackout still prevailed. However, a new form of radio intelligence came into play whose significance as a way of protecting convoys was even increased by the lack of Ultra.

By the fall of 1942, most Allied escort groups had at least one vessel equipped with an automatic high-frequency direction finder.[86] This meant that German convoy-sighting transmissions from U-boats were often picked up directly by HF/DF-equipped escorts. At this point antagonists would be no more than thirty miles apart and escorts could be dispatched down the bearing to the U-boat position. Approaching escorts forced the submarine to submerge and lose contact, which would give the convoy a chance to escape. Detached escorts armed with radar and sonar, moreover, could often gain contact with the HF/DF-located U-boat, and when they got close enough they could attack it directly. In this way close-up defense of convoys was greatly improved even when the U-boats were already in contact. Shipborne HF/DF had a significant part to play in neutralizing the U-boat tactically and also in making it feasible for the escorts to fight back aggressively.

Either side might have won the Atlantic War, and in the end there was no cosmic development to separate winner from loser. It was, rather, an accumulation of small, incremental advantages and, even more important, the speed of reaction to changes, that accounted for the Allied victory. One highly significant factor was the sudden improvement in the ability of Allied ASW forces to find U-boats when they were most dangerous, that is, at the point of attack. Shipborne high-frequency direction finding provided that advantage to the Allies. Because the Germans never suspected the cause of this ominous change, HF/DF never became a part of the spiral of development and counterdevelopment. Shipborne high-frequency direction finding was never subjected to its only really effective countermeasure: radio silence.

Dönitz may have been forced to consider the risk to his submarines from strategic direction finding, but he never acknowledged it as a tactical device, even when the physical evidence was clear. U-boat Captain Cremer has left a record of various early encounters with HF/DF-equipped escorts, and he notes his earnest speculation at the time as to what could have accounted for their success in locating his U-boat. On 17 August 1942, not long after sending off the mandatory convoy-sighting radio signal, Cremer first saw a "strange latticed contraption" (the characteristic birdcage-shaped Bellini-Tosi crossed-loop antenna of British HF/DF) on an Allied warship. Then he saw another. He was soon attacked by a Liberator bomber and destroyers, even though he had remained below the horizon and believed it highly unlikely that the convoy could have spotted him. Cremer remembers thinking that the masthead baskets must have had something to do with the attack on him, and yet he was sure that he could not have been detected by radar, about which he seems to have been well informed.[87]

At this and other points in his narrative, Cremer asserts that he and his colleagues never expected to encounter effective high-frequency direction finders afloat. Such myopia is hard to account for in hindsight; how can it be that reports of those strange antennas on enemy vessels went unremarked at U-boat headquarters? Especially since Cremer describes the activities of Professor Küpfmüller, the physicist who headed Dönitz's scientific-operations staff. Apparently Dr. Küpfmüller was present at all situation conferences and was active in soliciting input from operational commanders.[88]

Actually, this may have been part of the problem. While Dönitz and his staff were very close to their fighting men and were generally attuned to their perceptions of the sea war, what was lacking was an impartial, regular, clinical dissection of the incoming information by independent scientists not connected with the military hierarchy.[89] While Dönitz initiated periodic thorough investigations, his operations were not subjected to the steady, persistent analysis of prosaically accumulated detail to which the British and American operational research teams subjected their operations.[90]

Thus the link was not made between antennas and interception of U-boat transmissions. It has been shown that "at the latest, ever since the spring of 1943, the Germans had clear proof of the presence of shortwave D/F equipment on escorts."[91] This had been obtained by deciphering Allied radio messages. German agents even took photographs, from

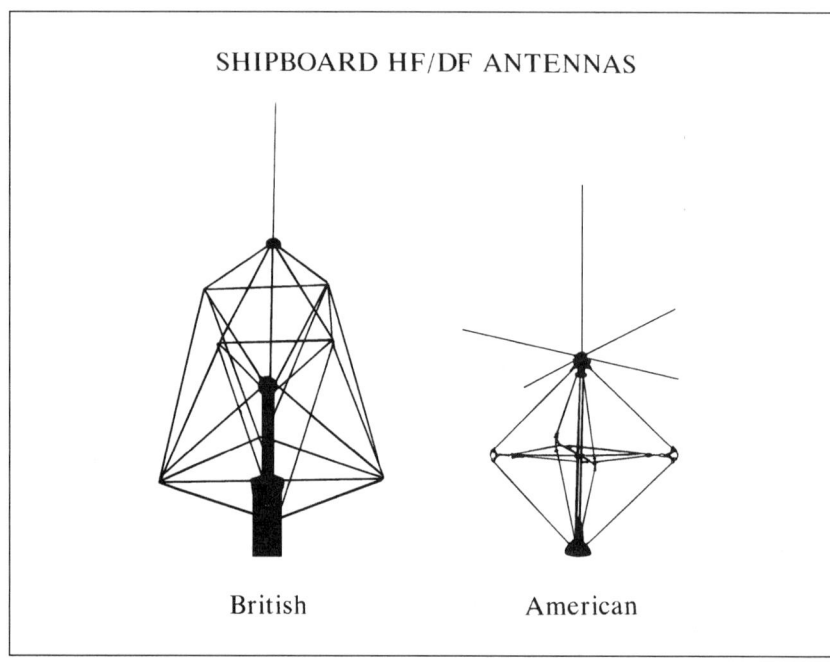

SHIPBOARD HF/DF ANTENNAS

British   American

Fig. 2.3. British and American shipboard HF/DF antennas.
Drawing by Roger G. B. Broome

a house near Algeciras in Spain, of British Huff Duff–equipped escorts at anchor at Gibraltar. But the censors misidentified the strange antennas as belonging to radar, a British device "about which we still have no precise information."[92]

Dönitz never received the kind of scientific advice with which the Allied naval leadership was inundated; in fact he was "left fighting his war by his fingertips with little scientific assistance."[93] That is how exaggerated fear of radar and at the same time ignorance of Huff Duff were perpetuated. "Defective technical and scientific thinking . . . and a good deal of arrogance" have been blamed for Dönitz's failure to accept the threat of direction finding.[94] BdU could neither imagine nor would seriously consider the possibility that the enemy could tune into the minimal signals, lasting only seconds, by which they exercised tactical control of their U-boats. This constituted a real failure of vision at the highest level. Dönitz himself must shoulder much of the blame for lack of scientific sophistication and for failing to build up an operations research structure.

Still, protecting a convoy in theory and doing so in practice were two different things. In practice, Dönitz's finest hour had not even arrived yet. While it is commonly agreed that convoying, especially escorted convoying, is the prime defensive measure against submarine attack, Dönitz was not wholly mistaken in his assessment of convoys as vulnerable concentrations of targets, at least potentially. Benjamin Brooks, who served on the old destroyer *Haraden* before it was sent to Britain in the Destroyers for Bases deal, recalls a later experience as first lieutenant in charge of the deck force on the destroyer *Belknap*.

> In late 1942 we were called upon to reinforce a convoy which had been under attack and suffered losses while en route from Trinidad to Recife, Brazil. It was a 6-knot formation bucking a 2-knot coastwise current. We patrolled at 12 knots for maximum maneuverability and to assure reaching optimum attack speed as quickly as possible in the event of a contact. We found we could keep station in our patrol sector by simply reversing course, steaming at right angles to the convoy, and allowing it to catch up through our transfer from one heading to the next. . . . That was the essence of the sound screen [a screen formed of escorts equipped with echo-sounding gear]. It was developed not by logical developments, but by the number of escorts available. Sometimes we'd be lucky to have three in a situation that required seven or eight. And there were a lot of penetrations [by U-boats], and in the early days the convoy would be decimated.[95]

Later, with more escorts available, new tactics could be employed. As Brooks explains:

> Under those circumstances it was customary for one escort to take a nighttime position astern to counter a possible surface attack from that sector. Heretofore, the screen had not been large enough to do this. Now, unknown to the convoy, *Belknap* was its first roving tailback. As we approached the formation, a Portuguese ship in the rear took us under fire. Tracers started to fly over the bridge, and we stole back into the night. The next morning we brought a Portuguese-speaking cook to the bridge and hailed the ship. Some of what he shouted was plainly not very nice, and we came to an impasse. Finally, they displayed a blackboard on which was chalked in English: "At 2130 hours last night we fired at one small submarine."[96]

In November 1942, Dönitz was directed to divert U-boats to cover Operation Torch, the Allied landings in North Africa. Yet thanks to a concentrated building program, Dönitz now had the number of boats with which he had wanted to start the war. He could keep almost one hundred boats operational while an equal number would be in transit and the same number again in port being repaired and refitted.[97] In the same month as Torch, more than 700,000 tons of Allied shipping were sunk. This had been Dönitz's original monthly goal, which he had estimated as necessary to starve Britain out of the war.[98]

In February and March 1943 it became increasingly difficult for the Allied merchant convoys to maneuver past the growing number of U-boats driving across the Atlantic in their extended patrol lines like multipronged harvesters poised to mow down everything in their path. In two weeks in March, four eastbound convoys, SC 121, HX 228, SC 122, and HX 229, lost up to 20 percent of their ships (including the *Southern Princess*) in huge convoy battles. And though this "Bloody Winter" saw the worst weather in fifty years, German successes continued to mount until March 1943, when victory appeared imminent for the U-boats.[99]

This was to prove an illusion. It had taken too long to build up the U-boat forces. Just as they reached the peak of their numbers and powers, so did the Allied ability to strike back.

Shipborne HF/DF was just starting to get into its stride during a series of German attacks on the sixty-one-ship convoy SC 118 in February 1943. The twenty U-boats engaged (which had been accurately positioned for interception of the convoy by B-Dienst decryptions) made 108 transmissions during a period of seventy-two hours. Each one of those transmissions represented an opportunity for the Allies to pinpoint the position of the U-boat making it, and of the three submarines sunk during the running battles, at least two were located by HF/DF. One of the submarines, the U-624, was actually caught on the surface while transmitting an account of the previous night's attacks.[100]

The following month, a report on "Measures for Combatting the Submarine Menace," issued by the Combined (Allied) Staff Planners, enigmatically states: "The radio traffic involved in the operational control of U-boats at sea by their H.Q. ashore enables our intelligence organization to forecast probable U-boat intentions."[101] This subtle suggestion of codebreaking, however, misses the point that while Ultra could often warn of planned U-boat concentrations, sometimes confrontations took place anyway.

In the case of SC 118, the B-Dienst information enabled BdU to position his patrol line so accurately that the convoy plowed straight into the middle of it. At this point Ultra could be of no further assistance, and it was shipborne HF/DF that helped actively to defend the convoy and that also guided escorts and aircraft to successful U-boat sinkings.[102] As usual, though, no progress was made in isolation. In addition to an increase in numbers of escorts of all types, by early 1943 the Allies had also deployed centimetric (S-band) radar. This was a big improvement on the previous metric radar, as it was much more sensitive and was capable of detecting an object as small as a U-boat's periscope.[103]

Most significant, however, was the fact that the Allies at this time also came up with a new strategy. The information from radio intelligence, including Ultra decrypts and HF/DF fixes, was to be used in a new way as of 20 March. Up to this time, intelligence information had mostly been used to direct convoys away from concentrations of U-boats. In spite of the slowly growing number of escorts, whenever the new large wolf packs found convoys the engagement usually proved disastrous for the merchantmen. But an Atlantic Convoy Conference took place in March, and this meeting of British, Canadian, and American officials sharpened the Allies' determination to use all resources to defeat the U-boat menace. The still-sparse support groups, as well as all available aircraft, were from now on to be thrown into the fight more aggressively than ever. Ultra information would be used to summon up these resources to speed to convoys in danger of attack and to "fight the convoys through."[104]

Another important result of the conference was that sufficient allocations of air support finally allowed the Allies to begin to close the Air Gap, starting with patrols by land-based VLR aircraft. A line of U-boats steaming fast on the surface was particularly vulnerable to VLR B-24s. The Air Gap was also tackled that spring with the first use of British and American escort carriers in the defense of convoys.

These small carriers, referred to in the U.S. Navy as baby flattops (or jeep carriers), were mostly rebuilt or converted freighter hulls that carried about twenty planes. There were early criticisms of these carriers, which were seen as particularly ill suited to the North Atlantic because the bad weather meant there were many days when they could not get their planes off. Moreover, when they first came into service they tied up a number of scarce escorts, which were necessary for their defense. Bad weather also affected the escort screen by preventing at-sea refueling, which sometimes forced the whole carrier group to abandon its convoy.

The destroyer *Belknap* was one of the escorts with the USS *Bogue* on its first North Atlantic sortie in March. Captain Brooks (then the *Belknap*'s executive officer) remembers:

> Landing aircraft would one moment be heading straight at the *Bogue*'s propeller. The next it seemed they were flying into the wall of a handball court. These circumstances resulted in a series of crashes which broke the back of nearly every plane she carried. . . . Both [refueling and landing] efforts were abandoned for these reasons.[105]

Still, for the first time it became too costly for the Germans to attack convoys. In spite of the drawbacks of the escort carriers, the new strategy succeeded: using radio-intelligence information to send support groups racing to the assistance of threatened convoys where they would meet the U-boats head-on instead of trying to avoid them. The problem for the U-boats became not how to find convoys—this their own radio intelligence had largely solved—but how to sink ships without being sunk.

Perhaps Admiral King was a bit premature when he wrote in August 1943: "The small escort, or support, carrier teamed with appropriate surface units defeated the German subs."[106] But there is no doubt that the carrier groups, with British, American, and Canadian components, helped to win the war against the U-boat. And one of the key reasons for this was that they were equipped with shipborne HF/DF.

In order to be effective, the support or hunter-killer groups had to have access to intelligence about the suspected presence of U-boats or their approximate locations. Early in the war it had proved futile to search wide areas of the ocean in the hopes of finding an enemy submarine. Intercepted German radio traffic provided invaluable Ultra information on the U-boats' whereabouts.[107] And shipborne HF/DF fixes from radio intercepts could indicate even more current submarine locations.

Since intelligence gleaned from codebreaking was only useful so long as the successful break-in was unsuspected, the location information had to be disseminated to escort groups indirectly, always covered by other information. HF/DF was frequently used as a cover because its existence (though not its full potential) in the form of large antennas could not be hidden. It is not easy, nor maybe even possible, to reconstruct exactly what part each source of information played in each successful anti-U-boat operation.[108] Eventually, however, the *Bogue*'s aircraft sank nine U-boats and one Japanese submarine, which established a record for the war. HF/DF can be shown to have played a vital part in at least three of these sinkings.[109]

When Dönitz's U-boats had had spectacular success against eastbound convoys in March 1943, British authorities had given serious consideration to abandoning the convoy system. Instead, the support groups had been formed, which in April began to turn the tide. And when the U-boats attempted a repeat performance of the March successes early in May, they were driven away from the slow, westbound convoy ONS 5 with a loss of ten U-boats altogether, in return for only twelve merchant ships sunk.[110] This was not a sustainable ratio for Dönitz and is often considered the turning point in the Battle of the Atlantic. From then on the U-boats were on the defensive.

Now the Allies had really struck back. Not only did they employ new tactics, they also brought together new weapons such as ahead-throwing mortars like Hedgehog and Mousetrap (and later also Squid) and Fido, an acoustic torpedo, the new submarine-detection device (shipborne HF/DF), more escort vessels, carrier-based aircraft, and long-range, land-based bombers. The Germans lost thirty-seven submarines in May alone—Black May, as Captain Cremer called it. On the 24th of that month, just eight weeks after the introduction of the new aggressive Allied strategy of fighting the convoys through, Dönitz withdrew his boats from the North Atlantic, sending them to operate in southern waters.[111] There many of them ultimately fell prey to the American escort carriers that were to prove successful as the core of U.S. hunter-killer groups.[112]

That summer the British concentrated their ASW efforts in the Bay of Biscay transit area once again, while the Americans tackled the U-boats in the central Atlantic.[113] The hunter-killer groups sailing out of U.S. East Coast ports in the summer of 1943 were equipped with Henri Busignies's shipboard HF/DF and were able to find and attack Dönitz's U-boats in mid-ocean. The escort carrier planes made surface operations so dangerous that the U-boats had to remain submerged during daylight, radically reducing their effectiveness. They accomplished little in the central and southern Atlantic and their losses were great, including four milch cows.[114]

HF/DF was involved in these attacks, and May to September 1943 was instrumental in at least seven confirmed U-boat kills. In June, for example, planes from the escort carrier *Bogue* sank the U-118, a U-tanker, as it approached an ocean rendezvous with another U-boat. An HF/DF fix had guided the aircraft to the spot.[115] In the fall, Dönitz sent the survivors against the North Atlantic convoys once again.

This time U-boats were primed to counteract the aggressive convoy-defense strategy of the Allies. They were protected by a new search receiver—the Hagenuk—and they were armed with a new weapon: the T5 acoustic homing torpedo, Zaunkönig, or Gnat, as the Allies called it. They also aimed at a different opponent: the escorts. In theory, the long-range acoustic torpedo would enable the U-boats to attack escorts from a safe distance. Once the escorts were disposed of, the convoy would be vulnerable to attack. But Ultra information not only warned the Allies about the new torpedoes, decrypted U-boat reports about their efficacy helped in the speedy development of countermeasures such as the Foxer towed noisemakers. Also, by August, OP-20G, the American cryptanalytic section, had received new high-speed "bombes" (decryption equipment built on British models). In October and November these bombes had churned out decrypts so effectively that all efforts to concentrate U-boat wolf packs against Atlantic convoys were foiled, and in November OP-20G took over responsibility for American Triton decryption, exchanging results with the British.[116] Partly because Dönitz could exploit geography and shift his U-boats away from dangerous areas, he was not forced to concentrate on the technical aspects of the campaign. In fact it was not until the summer of 1943, when it was already obvious that the U-boats were facing possible defeat, that Dönitz consulted seriously with his scientists and engineers, finally looking to technology to save his campaign.[117] But by then it was too late. The Allies' own scientific establishment was already well organized for the war effort.

In October 1943, twenty-two U-boats were sunk attacking convoys, while their only victims were one naval escort and three merchantmen.[118] Gathering U-boats into packs had become too dangerous, and large group operations had to be dropped. This brilliant tactical innovation of Dönitz's had come close to cutting the supply lines to Britain but had been fatally weakened by reliance on radio. It was radio intelligence used against it that was largely responsible for the defeat of the wolf pack.

The occupation of the Azores in October gave the Allies control of the entire southern Atlantic with shore-based air. The escort carriers covered the central Atlantic, and the north became increasingly dominated by air patrols from Newfoundland, Iceland and Northern Ireland.

That winter there were virtually no merchantmen sunk as the U-boats were increasingly forced to remain submerged by Allied air cover.

Until summer 1944 the U-boats could no longer make any significant dent in the ever-increasing transatlantic flow of men and supplies. All Dönitz could do was to keep the Allied ASW forces occupied to prevent them from being used for offensive operations. Then in June Dönitz directed his U-boats to disrupt the Allied invasion of France, but hostile air and sea patrols made this impossible.[119]

In the final eighteen months of the Atlantic campaign, the U-boat use of radio was cut way back, which led to a corresponding drop in the importance of radio intelligence. Now the U-boats usually received instructions and sailing orders in writing before departure, instead of over the air after they had set out. Moreover, they were ordered to transmit only the most important tactical reports, even though the BdU log for 12 November 1943 admitted that they did not know what means of location of U-boats the Allies were using. Measures were also taken to make mandatory signals much harder to decrypt, and even to intercept. Signals were sent with coded references to the written orders, and in December they began to use separate keys for each U-boat. So few signals were sent that these proved to be very hard to crack.[120]

Finally, in addition to abbreviating the contents of their signals, the Germans also developed a technique for reducing transmission time. They created a burst or "squirt" transmitter, called Kurier, specifically designed to thwart direction finding by its ultra-high speed. The Telefunken Company in Berlin had been at work on this device since 1942. It was not a response to suspicions about shipborne D/F, but was directed at Allied HF/DF in general.

Kurier was operated by adding an attachment to the normal U-boat transmitter, and it sent a message previously recorded on magnetic tape. The message on the tape, which could contain as many as seven Morse letters, was compressed (or "squashed") for a transmission that could be accomplished in just 452 milliseconds. The effect was rather like running a record at 78 rpm that had been recorded at $33^{1}/_{3}$. Kurier had already been tested in some U-boats and it was thought that its widespread use would make interception almost impossible.[121]

After the war an engineer with ITT had the opportunity to inspect one of these Kurier devices that was on a captured U-boat in the Portsmouth (New Hampshire) Navy Yard. The German warrant officer who had been in charge of communications when the U-boat surrendered took considerable pride in demonstrating his Kurier to the American engineer. He showed how with a single lever he could obtain a Morse code dot, and with two levers he got a dash. By releasing a fixed lever he made a space, and with two levers a wider space. Once he had thus "printed out" his message on magnetic tape, the machine was set for transmission. The instruction manual for the Kurier gave the date, time, and frequency for each transmission. At the appointed hour a circuit breaker would close. After checking to be certain the time was synchronized with the ship's clock, the operator pressed a button that started the whole process.[122]

The German explained that this system allowed them to communicate with the home country in bursts lasting only fractions of a second. He was confident the enemy had never discovered this procedure. He was wrong, however. The Allies were aware of Kurier and they were concerned about it, but fortunately for them Kurier never came into general operational use. Beginning in August 1944, though Kurier was installed in an increasing number of U-boats, the system for its use was never really perfected.[123]

In this last phase of the campaign too, increasing numbers of U-boats had been fitted with the Schnorkel air-intake tube that had been invented in Holland before the war. The Schnorkel allowed U-boats to remain submerged most of the time, thus avoiding detection from the air. In this way the boats achieved some success in shallow coastal waters, where they were also safe from sonar as they blended in with sound-reflecting rocks and shoals.

But by September 1944 the U-boats' home ports in the Bay of Biscay were at risk of capture, so Dönitz ordered them to withdraw to Norway. This limited their range to the waters around Britain, where owing to their tendency to remain submerged for long periods (sometimes fifty and even seventy days), they had little impact.[124]

At this point the only hope for Dönitz was to completely change the nature of the battle. He could have done this earlier in the war by imposing radio silence on his U-boats, but he chose not to. Now the only change left that might have enough impact to affect the campaign was for Dönitz to bring out his new high-submerged-speed submarines that were already in production. But the new types XXI, XXIII and XXVI, which might have reactivated the German U-boat offensive once again, failed to become operational soon enough. This was at least partly due to damage inflicted by heavy Allied bombing attacks on building yards, but mostly it was because development had not been started soon enough.[125]

Until the summer of 1943, Dönitz had believed he could win the war with sufficient numbers of his prewar U-boat types. By then it was obvious that the U-boats had been forced from the surface, so they had to develop new types with high underwater speeds and endurance. Experiments had been carried out since the beginning of the war on just such a boat—the Walter—propelled by hydrogen peroxide. But time had run out, and priority was switched to the transitional types XXI and XXIII.[126] In fact, the war in Europe ended on 8 May 1945, only shortly after the first type XXI—the U-2511—had put to sea. The U-2511 had just had time to make a dummy attack on the British cruiser *Norfolk* without being discovered. Had the war continued longer, Dönitz once again might have had a submarine capable of evading Allied detection.[127]

So much in war depends on timing. Inevitably the vast industrial output of the United States eventually overwhelmed Germany's resources. By the end of 1942 the States was already producing over eight million tons of merchant shipping, and by 1945 it produced over fifty million tons.[128] At the same time escorts of all sorts, including carriers, were being turned out at an extraordinary rate.

The number of U-boats also increased during the war. Seventeen a month were being completed in 1942, for example, and by January 1943 Dönitz was able to keep one hundred boats at sea. But their individual and strategic effectiveness declined dramatically in the face of evergreater numbers of enemy merchant ships and escort vessels.[129]

More than anything else, however, this campaign was characterized on both sides by the persistent and double-edged use of radio. The U-boats brought upon themselves a large measure of their defeat by their continuing dependence on radio transmissions. Shipborne HF/DF, an invaluable defensive and offensive tactical tool, enabled the Allies to exploit this major weakness. But HF/DF was by no means alone. Sonar, radar, and most of all Ultra also played a major part in the defeat of Dönitz's U-boats.

CHAPTER 3

Sonar, Radar, and Ultra

O n 20 April 1943, Adm. Francis S. Low, assistant chief of staff (antisubmarine), wrote a memorandum for Admiral King, commander in chief of the U.S. Navy. Admiral Low acknowledged that a "correct evaluation" of the antisubmarine situation was hindered by "a common conception that there are panaceas in the form of new devices, most of which 'look promising.'"[1]

The three "devices" that promised most for the U.S. Navy in its struggle against the U-boats were sonar, radar, and the direction finder. It was in the field of electronics and, for two of the three devices, radio electronics, that "panaceas" were being sought.

From its earliest days the development of radio had been closely and fruitfully associated with naval warfare. Radio waves are generated by electrical power, and the difficulty of making portable power sources prevented land forces from harnessing radio effectively for decades. At sea, however, ship engine-room generators produced abundant electrical energy that was easily harnessed to the wireless. Indeed, naval radio communication, already used very effectively by the Japanese in the Russo-Japanese War of 1904–5, was well advanced by 1914.[2]

As quickly as radio waves were harnessed for naval communications, the ability to extract directional information from those waves (direction finding) was given a naval purpose. By 1918 radio direction finding was already an established navigational aid. It had also been used in Britain with some rudimentary success to detect the presence of German submarines through their radio transmissions. Direction finders were, indeed, the oldest of the "new" devices expected to solve the U-boat problem during World War II. Still, direction finding has remained little known to the general public, and its role has been even less understood.

Significantly, the Germans at the time blamed radar for many of the wartime successes that were actually achieved by Ultra and by shipborne

HF/DF. Beginning in 1942 and increasingly in 1943, the danger to U-boats of detection by radar was greatly overestimated by the German command, to such a degree that even apparently clear indications of Allied use of HF/DF were interpreted as evidence of radar. This blinded the experts as well as Dönitz's staff to any other explanation of successful evasive moves by convoys; especially to the unwanted explanations of codebreaking and direction finding.[3]

Dönitz was aware of HF/DF and yet he continued to rely on radio communication. The ambivalence of this position is clearly demonstrated in the following extract from spring 1944 of the history written by Dönitz's son-in-law and former staff officer, Günter Hessler:

> Prior to the last months of 1943, the U-boat commander had used every opportunity to [send] a situation report and this continual stream of incoming signals enabled the U-boat Command to maintain an up-to-date record of losses and their probable causes. Now, however, the efficiency of the enemy D/F organization [German radio-intercept information showed that each signal was picked up and ranged by an average of twenty D/F stations] deterred commanders from using their radio, some boats remaining completely silent for over six weeks on end, and without these vital situation reports it was almost impossible to make an accurate assessment of losses.[4]

When Dönitz wrote his memoirs in the 1950s using the study Hessler had prepared for the British Admiralty, he was finally able to rule out the possibility of spies. In the end he concurred with Hessler's conclusion that the means being used to trace his U-boats must have been not only the radar he had already blamed, but also probably HF/DF. At the time of writing neither Dönitz nor Hessler had had access to the still-classified information about Allied codebreaking.[5]

It has only been since 1974, with the first revelations about Ultra, that the British breaking of the Enigma ciphers has been recognized as the reason for much of the successful rerouting of convoys in the second half of 1941 and later. The Ultra secret had been kept so well until then that memoirs and other accounts of the Battle of the Atlantic were seriously skewed. Usually, like Dönitz's memoirs, they gave too much credit to radar for otherwise inexplicable Allied successes.

After the cessation of hostilities, in January 1946 the U.S. Navy, in conjunction with ITT, gave a thorough demonstration of the capabilities of American HF/DF for representatives of the press. Following that demonstration, detailed articles about the device and photographs of it appeared in metropolitan newspapers all over the country. Indeed, as late as the early 1970s, navy archivists continued the thirty-year practice of routinely fending off awkward inquiries regarding anti-U-boat successes by crediting HF/DF with more than it actually had accomplished. They did this in order to protect the classified source of information about U-boat movements: the Ultra decrypts.[6]

Yet even with such publicity, the real nature and contribution of HF/DF has remained obscure. A partial explanation for this may be that while the existence of the technology was widely known, its specific capabilities were shrouded in secrecy. Even if it was assumed that direction finding was possible on shortwave transmissions, it was not at all clear how long those transmissions had to last in order to provide a good fix. Indeed, the minimum message length that a high-frequency intercept system can handle is still undetermined today. World War II Kurier transmissions lasted 452 milliseconds, while in 1960 the navy could foresee the need to detect and fix transmissions as short as one millisecond in duration.[7]

Still, even the fastest direction finder cannot operate in the absence of radio transmissions, so it is vital to encourage the cooperation of the enemy by allowing them to believe that their transmissions are secure. In the postwar world that meant allowing the Soviets to believe they could squash their signals below any existing system's ability to recognize them. It is worth bearing in mind, therefore, that continuing postwar security concerns have possibly been responsible for helping to obscure the real record of HF/DF achievements in World War II. A brief survey of the development of the other major electronic devices and technologies in use during the war—mostly British, American, and German—and a detailed description of naval direction finding should help to clarify the record.

Between Pearl Harbor and the end of the war with Japan, the U.S. radio industry produced nearly twice as much radio- and radar-communications equipment as it had previously produced since commercial radio began, around 1922. This phenomenal increase during the Second World War was built upon the published and widely known scientific developments of the previous forty years.[8]

In October 1901 F. M. Barber, a retired U.S. Navy commander with some signals experience, was recalled to active duty in order to study

European progress in radio electronics. His reports laid the foundation for the U.S. Navy's electronics program.[9] After World War I a rather free exchange of scientific information, temporarily interrupted by the fighting, resumed in scholarly journals and in international conferences. Many of the practical developments then, as later, were the product of commercial firms. But this peacetime openness clashed with government notions of secrecy. Even the former Allies ceased the official exchange of scientific information they had undertaken during the war. In 1921 the United States declined a British offer of scientific cooperation, and that set the precedent for separate, government-sponsored research and development during most of the next two decades. Naturally the combination of scientific openness and governmental secrecy led to considerable duplication of effort and to hazy ideas about the authorship of new developments. This confusion marks the stories of sonar and radar and is especially apparent in the history of high-frequency direction finding.[10]

Many of the devices developed by the Allies for naval use in the Battle of the Atlantic were the result of the combined efforts of numbers of people in Britain, the Commonwealth, and the United States. As retired navy captain Benjamin T. Brooks notes: "If demand for technical research is strong enough in assorted areas the idea can have many fathers and many birthplaces. This is a phenomenon that makes patent lawyers happy."[11]

Sonar was developed during the First World War and improved upon afterward. The principle of its operation is simple. A transducer (which converts energy from one form to another) housed in a retractable dome fitted on the hull of a vessel acts like a loudspeaker to emit sound waves underwater. These then bounce back from an obstacle and are picked up again by the transducer. The directional waves sent out by sonar are generated by a quartz crystal oscillator transmitting regular electrical impulses that the transducer converts to underwater sound. The echo received back is converted by the transducer into a form an operator can perceive. It is then relayed to the control equipment, usually located on the bridge level, from where in World War II the ominous return from a contact was often broadcast directly on the bridge loudspeaker. Thus, upon hearing the sonar return, an experienced conning officer could almost instantaneously adjust his course and speed to counter the target's moves.[12]

By 1939 the British had perfected the essential features of asdic, but it continued to have substantial limits of range and reliability. Success in determining correctly whether a contact was really a submarine or just any number of other obstacles (which could cause a deceptively similar return) depended heavily on the skill of the asdic operator.

Similar early progress had been made independently in the United States. The Sound Division of the Naval Research Laboratory (NRL) had discovered the same techniques and had developed equivalent devices. The U.S. Navy, too, put excessive trust in this one antisubmarine device. As has been noted, "the interwar Navy had lavished talent and money on sonar but had virtually ignored improving weapons and equipment."[13] Continued progress during the course of the war extended the range and accuracy of sonar, rather than creating basic changes. Gradually it became possible for highly trained operators to ascertain not only the presence but also the direction of travel of a submarine, as well as its speed and depth.[14]

It is often asserted that during World War I, the German submarine campaign against Allied shipping very nearly succeeded in cutting off Britain from all seaborne supplies. And yet in spite of this promise of success, by 1939 Germany had built only fifty-seven U-boats. At the same time the Royal Navy had over one hundred destroyers fitted with asdic, as well as forty-five sloops and twenty or so trawlers ready for antisubmarine work. The Royal Navy felt matters were well under control. It had such confidence in asdic, in fact, that between the wars it did not conduct a single exercise based on the theory of protecting convoys against submarines. Should there be another war, asdic would do the job.[15]

To be sure, when war really came asdic did account for many U-boat sinkings in the Atlantic, but it was by no means infallible. Asdic had substantial weaknesses and limitations, as did each of the other antisubmarine devices. Not the least of these weaknesses was the "active" nature of asdic. The underwater sound waves transmitted by the equipment could give away the presence, even the location, of an antisubmarine vessel, and so lay it open to attack by the very submarines it was hunting.

U.S. Navy destroyer officer Ben Brooks, who had extensive experience in the Atlantic war, remembers vividly the difficulties of running down a U-boat using sonar: "Your assured sonar range up in the North Atlantic, for all depths, was about 1,200 yards—a little better than half a mile, and that's not very far. You're steaming at fifteen knots—a mile in four minutes—and that means that if you pick up a target you often have less than two minutes to get over it and drop your charges."[16]

Sonar's short range kept escort crews in a constant state of unrest. Brooks notes the routine when a contact was made:

> You wouldn't get him until he was about 1,200 to 1,500 yards away and that meant that first you conduct an urgent attack, and get down and just let him know that you know he's there, and mix up the sauerkraut a little bit. We always had a third of our armament manned and ready to go at all times, and we would conduct that first urgent attack with the watch that was on duty. And we wouldn't try to change it, wouldn't upset the proceedings by bringing the first crew in so they could run the attack; there wasn't time. So after that first attack, then you'd hit the general alarm.
>
> As a matter of fact the cables to the depth charges went through a piping through the mess deck, where the guys would be eating and writing letters and so on. The first thing they'd hear of a contact was the cable moving, and they'd yell, "get going," and they'd be halfway out on deck and running to their battle stations before we hit the alarm and before the first charges went off. It was a pretty fast operation. Then we'd close up with the first team: our finest sonar operators and conning officers. And our guns that had been unmanned would be manned. And that's the way we went to our battle stations in a combat antisubmarine situation.[17]

To add to the tension and confusion of antisubmarine duty, sonar echoes were not necessarily only returned by submarines. A U-boat's wake could reflect the sound waves, as could the sea bottom in shallow areas, as could shoals of fish. Even water currents of different density or temperature, the so-called thermal gradients, could reflect sound pulses. Brooks explains that "what we were looking for was a metallic 'tuck' sound from a possible sub." Subsequent pings could establish the target's width, and variations in the pitch of the echo indicated relative movement. A "teek," according to Brooks, indicated something approaching; a "took," movement away. "Fish, wakes and suchlike give typically mushy returns."[18] Because it took an experienced operator to identify the return as a genuine target, there were many false alarms. At least one reef was rammed in the mistaken belief that it was an enemy submarine.[19]

Tales of mistaken attacks on whales also abound in ASW mythology. Even with modern asdic this is apparently a continuing hazard. Britain's nature-loving Prince Philip warned the Royal Navy task force to leave the whales alone as it set off for the Falklands campaign in 1982. And Brooks's unfortunate encounter with a whale, when he commanded the *Belknap,* was no myth.

> Now we could usually identify the characteristics of a whale, except when a whale makes like a submarine—when it unaccountably follows standard U-boat attack procedure! A whale would be expected to change course, or do something, not follow a steady, logical attack course. So we went over and we dropped. We dropped a full pattern [of depth charges] on this thing, and it looked beautiful. We had a very excitable Greek [sonar] officer named Panagos. He said, "call the carrier and tell her we made a good drop and would like air support." And in nothing flat I had airplanes flying around our bridge and our mast like mosquitos around a jar of honey. And we sensed that this guy was coming up, and we could see oil on the surface, yellow oil.
>
> We circled around for our attack and the guys were all set up with gun shells in their hands ready to ram in, and to keep the loading lines busy. It was a very dramatic moment. Then there was a whoosh, and blood all over the place. At our reunions they never let me hear the end of that. I'm called Ben the Whale-Banger.[20]

During the course of the Atlantic campaign, Admiral Dönitz was able to implement his tactics of night surface attacks against convoys, which proved to be the most successful technique for foiling sonar. German naval scientists also developed a number of other antidotes, among them Pillenwerfer, capsules that released a cloud of gas bubbles into the water, reflecting back an echo to a sonar wave. But sonar operators quickly learned to discount the totally stationary signals received from the Pillenwerfer: unless resting on the sea bottom, a submarine must maintain some forward motion for balance.

Early in the war, efforts had been made to cover U-boats with rubber. Later a special reflective latex coating (called Alberich for the dwarf of German legend with powers of invisibility) was created so that submarines would absorb rather than bounce back the sonar waves. But none of the coatings were used very much. The rubber soon wore off when exposed to the harsh conditions of the Atlantic, and the latex was too costly and difficult to apply to be used extensively.[21]

The struggle to stay ahead in the battle between technical devices

Benjamin T. Brooks, skipper of USS *Belknap*, between San Juan and New York, 1943.
Courtesy of Capt. Benjamin T. Brooks

and their countermeasures has been characterized as "symmetrical" in the case of sonar.[22] This means that a new development on one side was closely followed by a new countermeasure on the other, and so forth, in an unending dance. The struggle was also symmetrical and even more pronounced with regard to antisubmarine radar.

The principle of radar is similar to that of sonar except that radar transmits radio waves through the air instead of sound waves underwater. A radar transmitter works by emitting brief intermittent pulses. A receiver picks up the return echo when the wave encounters and bounces off an obstacle. Since radio waves travel at a constant speed of 186,000 miles per second, the distance of the obstacle can be computed by the delay between the pulse transmission and its return.[23]

There were several problems that had to be overcome in order to make radar an effective detection device. While the earliest radar could indicate the distance of an object, direction and bearing were not clear. Available power sources produced only long waves, yet the longer the wavelength transmitted, the larger the antenna necessary to send and receive the return wave. With a longer wavelength, too, only objects of considerable size were detectable. For radar to be useful militarily it was necessary to produce directional antennas, so that wave transmissions could be aimed in a particular direction. Only then would the direction of return be known. It was also necessary to reduce the size of the waves transmitted, so that antenna could be mounted on aircraft and naval vessels and smaller objects could be detected.[24]

By the mid 1930s, when radar development really began, radio-wave theory was already well established. Radio waves may be measured either according to their size, that is the length of the wave (say from crest to crest), or according to their frequency—the number of the wave cycles per second. Frequency and wavelength are inversely related so that the shorter the wave, the higher the frequency. A three-hundred-meter wavelength, for example, travels at one million cycles per second. That is expressed as either one megacycle (Mc), or one megahertz (MHz).[25]

Of the four nations that developed radar independently in the 1930s—France, Germany, Britain, and the United States—the British were the last to do so. However, by the outbreak of World War II they had made greater progress than the rest. In 1935 the Scottish scientist Dr. Robert Watson-Watt was the superintendent of the radio department of Britain's National Physical Laboratory. Short, stout, bespectacled and brilliant, Watson-Watt began work with his team of experts to develop equipment that could make practical use of the research they had been doing on radio waves.

They had noticed that such waves, particularly those in the higher frequencies, were returned as echoes when striking certain obstacles. The challenge was to make a device powerful enough to transmit a radio pulse to a considerable distance, yet sensitive enough to receive the return wave. Watson-Watt accomplished this by modifying a goniometer, a device that had been used in the radio direction finding (RDF) field since early in the century. It could handle the very high frequencies necessary. He also deliberately used the well-known initials RDF for his new device in order to preserve the secrecy of the project. This led to considerable confusion of terms, which has persisted until today.[26]

Thanks to the early success of his work, Dr. Watson-Watt has become popularly known as the inventor of radar. Indeed, in his autobiography he refers to radar as "what I invented."[27] And later on, "unblushingly," he sums up his accomplishment by quoting the Duke of Wellington: "By God! I do not think it would have been done if I had not been there!"[28]

Destroyer *Belknap* on hunter--killer operations north of the Azores, 1943. *Courtesy of Capt. Benjamin T. Brooks*

Certainly Watson-Watt was largely responsible for the promotion and practical development of radar in Britain, and the scientific and engineering accomplishments

were the result of his direction of a fine team effort. Radar was, however, a development involving advances in several associated fields, and
it was being worked on simultaneously in a number of different countries.[29]

In the United States, early experiments by the Naval Research Laboratory on radar had not been followed up aggressively because it was
hard to attract funds for research and development when no urgent operational need was apparent. By 1935, however, according to the NRL
war history, American radar "had been conceived, developed, and was
ready for the pilot-model stage."[30]

When Capt. H. M. Cooley was director of the Naval Research Laboratory (1936–39), he was particularly skilled at securing funds for new
technologies from skeptical bureaucrats and congressmen. Drawing on
an unexpected flair for showmanship, Captain Cooley mounted many
impressive demonstrations, convincing lawmakers of the practical applications of sonar and radar. Later Adm. Chester Nimitz even claimed
that it was in the mid-1930s, during Captain Cooley's directorship, that
NRL scientist Dr. A. Hoyt Taylor "discovered the phenomenon of radar"
while he was involved in research on high frequencies.[31]

In the mid-1930s it was already generally understood that the size
and configuration of antennas was crucial to the transmission and reception of radio waves. The basic antenna, a metal rod known as a dipole, transmits a signal in all directions. When two dipoles are used together side by side, the signal shape becomes elliptical, and the more
dipoles that are used, the narrower and more directional the ellipse becomes. With a reflector, the signal direction can be controlled even more.

In addition, scientists knew that the length of the dipole needed to
be half the length of the wave to be transmitted. In early experiments
with radar they found that the objects detected by the radar pulses had
to be at least half the length of the transmitted wave, or the wave simply
washed over the object without bouncing an echo back to the receiver.
Thus the shorter the wavelength (that is, the higher the frequency), the
smaller the antenna needed and the smaller the detectable object.[32]

Throughout the 1930s and continuing through the war years, major
radar research was involved in the effort to improve directional antennas, to narrow or concentrate the wave beam by raising the frequency
used, and to increase the power output that would make a narrow beam
possible. At first, large land-based radar was developed using vast an-

tenna systems. Huge transmitters broadcast signals whose wavelengths were measured in meters.

But radar for warships (especially small escort types) and aircraft needed to be much smaller, for reasons of economies of weight and space. Eventually many different types of naval radar were produced with such varied tasks as air and surface warning; gunnery direction; and identification, friend or foe (IFF). The major focus here is on antisubmarine radar, especially those sets suitable for mounting on escorts and aircraft.[33]

The story can be confusing, because several different types of radar were being developed and produced at the same time. To complicate matters, because there were never enough new and improved devices to replace the less-effective ones, many different sorts of radar were in use simultaneously.

In 1939 the 7½–meter radar, type 79, was already installed in a handful of large Royal Navy ships. These first shipborne radars were designed not for submarine detection but for air search, and they used a large "bedspring"-type antenna. While these radars were completely directional, they were not effective in picking up buoys or small surface objects because of their long wavelength. The most common British radar for escort use, introduced early in 1941, was RDF 286. With its tall mast and heavy fixed antenna, it was initially prone to weather damage and was of limited use for antisubmarine work because of the confusion caused by its side and back echoes. Later modifications brought some improvement in the device, but its most effective use was for station-keeping.[34]

The real breakthrough for naval radar had come when British scientists had developed the small and potent cavity magnetron power enhancer, capable of generating centimetric waves. A beam narrow or concentrated enough for submarine detection was being developed in Britain in 1940, but RAF night fighters as well as Bomber Command had first call on those devices actually produced.

The first effective surface warning centimetric radar for the Royal Navy was the 10cm type 271 with rotatable antenna, narrow beam, and a range of 6 to 7½ miles on a destroyer. The 271, which was capable of picking up a submarine's periscope, was installed in some Royal Navy vessels upon its initial introduction in May 1941, but because of Britain's limited manufacturing capability the supply was frustratingly slow.[35]

Capt. G. E. Creasey, director of antisubmarine warfare at the Admiralty, noted in a letter of 11 September 1941 to the superintendent of the Admiralty Signal Establishment:

> I cannot help pointing out the large number of escort vessels engaged in the Battle of the Atlantic and what a small proportion of those are now fitted with type 271. The point is, of course, that the Battle is already raging. The fact that ships will be fitted at some future date is something to look foreward [*sic*] to, but is no help at the present time when we are faced with a desperate struggle to get our shipping through.[36]

In October 1941 the Royal Canadian Navy and Royal Navy escorts protecting the fifty-two-ship convoy SC 48, as well as the Free French forces and U.S. Navy vessels that joined them, took a terrible drubbing at the hands of a U-boat wolf pack—they still had no radar at all. Convoy SC 48 was the one the USS *Kearny* was attempting to assist when it was hit and damaged by U-568 on 17 October, taking the first American fatalities of the war.[37] The *Kearny* was one of the twenty-seven vessels of Rear Adm. Arthur LeR. Bristol's support force. While all twenty-seven had sonar, only one in five of the vessels had been fitted with radar by the end of the year.[38]

Having foreseen the potential problems caused by the radar-production bottleneck, the Tizard Mission, a British scientific delegation, had brought the design of the cavity magnetron to the United States in the fall of 1940. Here two naval officers, Lt. Comdr. (later Rear Adm.) Frederick R. Furth and Lt. Comdr. S. M. Tucker, had already coined the acronym for radar. On 19 November 1940 the term was officially adopted.[39] Together, Furth and Tucker had been responsible for the original procurement of the model XAF air-search set, the U.S. Navy's first operational radar equipment. Now, working with the British cavity magnetron, large-scale production of centimetric (microwave) radar was undertaken at a rapid rate. By March 1941 type SG, a new 10cm set, had started trials, and it went into production by that summer. As a result, U.S. Navy vessels were generally equipped with centimetric radar earlier than were those of the Royal Navy.[40]

Still, there was an inevitable time lag between first production of any device and its widespread use. There has been much comment that the best and most up-to-date American ships and equipment went to the Pacific. But almost two years after the Tizard Mission, in August

1942, at the Battle of Savo Island off Guadalcanal, both the Royal Australian and the U.S. Navy vessels were still equipped with metric radar; only one light cruiser had the "new" centimetric radar. Back in the Atlantic it was not until 1943 that the *Belknap*, for example, had a surface search unit that was reasonably accurate in determining ranges on small targets like U-boat periscopes. Indeed, some of the metric surface search radars, like the American SC sets, remained in service until the end of the war.[41]

Like sonar, radar never became the panacea that had been predicted. For one thing, what the Allies needed in the Battle of the Atlantic was the ability to locate U-boats, but radar could only detect them on the surface. A U-boat that submerged was invisible to the very high-frequency radar waves, which could not penetrate the water. The prime use, therefore, for antisubmarine radar was to detect surfaced boats, though early radars did not outrange normal vision except at night and in poor visibility.

Radar "blips" and "pips," moreover, could be as deceptive as sonar "pings." In the air, birds could produce a radar reflection, and groups of sea gulls sitting on the water were known to have been mistaken for a sub contact. Surface reception was affected by wave action as well as by "back echoes" from nearby friendly vessels, and aircraft radar was prone to disappearing contacts, possibly caused by clouds or other aircraft. Even vagaries of the weather could affect radar.[42]

In November 1944, a U.S. Navy plane equipped with searchlight and "sniffer" (air ASW gear for detecting submarine gas exhausts) made a run on a radar contact that turned out to have been generated by a storm. The follow-up action report notes somewhat plaintively that "in addition to having a characteristic which can present a radar blip, this particular squall also tripped 'sniffer' and automatically released all the depth charges."[43]

Extensive training coupled with experience was necessary to produce an effective radar operator, and such people were hard to come by in wartime. Facing chronic shortages of radar personnel, the British Admiralty turned to the Canadian government for help. The first twenty Canadian physics graduates arrived in England for radar training in May 1940, and eventually Canadian radar officers were manning those critical posts in a large number of British warships.[44] The U.S. Navy would face a similar shortage of HF/DF operators.

Radar also had another weakness; like sonar, it was an active technology. Its very own transmissions might be picked up by enemy receivers. If a U-boat knew when it was being bombarded by radar waves, it only had to submerge to avoid them.

As with all of these technical devices, the Germans too had been working on radar. Funkmessortungsgerät, or Funkmess—radar—was in use in some heavy German ships when the war began. The Allies had what amounted to substantial proof of this in photographs of the *Admiral Graf Spee* after the ship was scuttled off Montevideo in December 1939.[45] So in order not to stampede the Germans into a search for radar detectors, every effort was made to preserve the secrecy of the effectiveness of radar. Inevitably, however, as soon as it came into extensive use it increased the likelihood of discovery.

In August 1941, the Germans were apparently not yet certain about the Allied use of radar. Hessler notes that "the only course was to order the U-boats to look out for signs of radar equipment and to report all their observations, which were passed on to the specialists who had the task of evolving countermeasures."[46] The result of such observations, in spite of strenuous security efforts on both sides, was an intense spiral of mutually stimulating development. This was concentrated in the German effort to detect the approach of a radar-equipped vessel or plane by search receivers such the Metox and later the Hagenuk, the Naxos, and finally the Tunis. The Allied countereffort was directed toward the development of new radar types that were impervious to each new German detector.[47]

While each step in the development of more sensitive and accurate radar devices hindered detection, the big leap came in the spring of 1941, when the first centimetric radars came into action. The German Metox could pick up metric radar, but as the Allies refined and narrowed the beam, this not only increased the efficacy of the device, it also made it impervious to existing search receivers. The new British centimetric 271 had the Germans stumped, and they even feared that their own search receivers were giving off detectable emissions, accounting for increased British successes.

Consequently use of the Metox was banned until German scientists came up with the Naxos search receiver, which was capable of picking up centimetric waves.[48] Other countermeasures considered or attempted included devices to jam Allied anti–surface vessel (ASV) radar and, like

Alberich, some submarine coating material that would absorb radar waves.

During the course of the war the development of centimetric radar was truly a joint Anglo-American effort, with the American contribution concentrated in the work of the Radiation Laboratory at MIT in Cambridge. Within two months of the Tizard Mission's gift of the cavity magnetron in September 1940, the "Rad Lab" had produced a microwave radar set that scanned the Boston skyline across the Charles River from MIT. Within two more months this device was being tested in an air-force bomber. Powered by the British cavity magnetron, the new radar had a Westinghouse pulse generator, a Sperry antenna, a Bell receiver, an RCA identification unit, and a cathode-ray display screen from General Electric.[49] American industry churned out vast numbers of devices of all types and for all services and uses—air, sea, and land.

Among the Allies, awareness of security issues surrounding technical devices was intense, and as late as August 1943, Secretary of the Navy James Forrestal had addressed a "basegram" to ALNAV directing that "under no circumstances shall reference to radar be included in any release of information in speeches or other public communication."[50] But it now appears that only the American public was ignorant of radar. Just one month after Secretary Forrestal's directive, a U.S. escort-carrier group in pursuit of a U-boat on the U.S.-Gibraltar route could not use the ship's radar nor the radar on the planes because the frequencies were thought to have been compromised. German search receivers were believed to be turned on and tuned in.[51]

Like other technical devices used by the Allies, radar was continuously under development during the war. That development was going on in several Allied countries at once, and there was extensive interchange of information among them. The many different models and types of radar had many different parents. Some British advances, especially the cavity magnetron, were largely developed and produced in the United States. Models made in the States for use on Royal Navy and Royal Canadian Navy vessels generally went by different names from essentially the same models produced for the U.S. Navy.[52] To add to the confusion, Britain continued to refer to radar as radiolocation and RDF until 1943, when the term *radar* was finally adopted by international agreement.

Then, too, the Allies mounted the sets differently. The British 271s and later 272s were installed on top of the bridge, while the American destroyers and destroyer escorts had their surface scanning sets, SCs and SGs for example, at the mast top. This higher position increased the range of the U.S. radar against the low silhouette of a surfaced U-boat. While in theory both the 271 series and the 10cm SGs were capable of detecting a U-boat's periscope, in fact it proved very difficult for any radar to get an echo from an object that small in the North Atlantic because of wave interference.[53] The Americans had an advantage in this regard when they took over responsibility for operations in the somewhat calmer waters of the central Atlantic area in 1943.

Contemporary assessments of the significance of radar often rated it above HF/DF. Capt. Donald MacIntyre, RN, was still convinced after the war that radar was the most effective device to prevent U-boat attacks on convoys, even though he was one of the most successful British commanders in the use of shipborne Huff Duff.

On 13 March 1943, for example, then-Commander MacIntyre, with Escort Group B2, successfully frustrated an aggressive wolf pack from making any attack on convoy ON 170 by his very efficient use of HF/DF.[54] And in May Commander MacIntyre, again with Escort Group B2, used shipborne HF/DF effectively to protect the twenty-six-ship convoy SC 129 from a dramatic confrontation with the U-boats.[55]

It was of this engagement that Dönitz wrote in his war diary: "Since detection on this scale and with such promptitude has hitherto been unknown, the possibility that the enemy is using a new and efficient type of locating device cannot be ruled out."[56] MacIntyre's comment on this quotation is significant: "It was true, of course, that the existence of an efficient, shipborne HF/DF was not known to the Germans. The submarines consequently had no inhibitions about use of radio, and each transmission was apt to result in the appearance from over the horizon of an aircraft or a surface escort accurately directed to the transmitter."[57]

Radar was not nearly as effective in these operations. In the end it seems pretty clear, however, that while the Allies may have had an edge on the Germans in the speed and the efficacy of their scientific and technical innovations, they produced no single war-winning device. There was no panacea. Still, each improvement in radar made it significantly more difficult for Dönitz's submarines to avoid detection, and his awareness of each improvement was a drain on him because it imposed the cost of seeking a countermeasure. Moreover, it can be argued that the success of the British campaign against the U-boats in the Bay of Biscay shows that a succession of even short-lived advantages can be as effective as a permanent advantage.[58]

Radio-electronic technology also had many other uses in addition to radar in the Allied anti-U-boat campaign. The development of the VHF radio telephone, for example, which was available by the end of 1940, had made possible direct and secure communication within a convoy-escort group and greatly improved the tactical control of the group. But far more important for the successful prosecution of the war was the penetration of the German cipher system.

British cryptographic efforts (the output of which was Ultra) focused on German military communications messages. These were enciphered on the highly complex, multiple-rotor, electromechanical machine with a typewriter keyboard known as Enigma. It had originally been developed in the 1920s for commercial correspondence, and, ironically, the early German patents had been filed not only in Germany but in several other countries, including Britain. In fact the theory on which the Enigma machine was based had been common knowledge in scientific circles since the nineteenth century, although there was not then the practical technology to actually build such a device.

The German Enigma D was sold to many countries between the wars, and both the British and the American military used it to develop comparable machines resulting in their own Typex and the electronic countermeasures. But this did not make messages enciphered with Enigma any easier to read, because the essential scrupulously guarded secret was the way in which the rotors were to be set each day.[59]

Throughout the war the Germans continued to believe, erroneously, that their codes were secure. As with direction finding, the Germans thought that their mastery of a known technology sufficed to give them an accurate assessment of its potential. But in fact, on 22 May 1940 the British intelligence organization at Bletchley Park regularly began to read the Luftwaffe's Enigma key. The naval codes were substantially more difficult to decipher, but this too was eventually accomplished.[60]

Still, cryptanalysis remained an imperfect tool. Even at its most successful, Ultra could only provide sporadic readings, and the lag time in decoding (anything from hours to days or weeks) meant that information was often not available in real time, hence in time for effective action. Also cryptanalysis, like all forms of intelligence, was only as effective a weapon as the operational use to which it was put. It took several war years to establish the mechanisms necessary to cull intelligence, to sort and analyze it, and to relay it in useful form to the appropriate operational commanders.[61]

Nevertheless, from June 1941 until January 1942 and again from December 1942 until the end of the war in Europe, Ultra was the most important method the Allies had of locating German U-boat groups. During these periods Ultra provided the British Operational Intelligence Centre with the BdU's radio orders positioning his U-boat groups in time to reroute convoys. The normal one-to-four-day decryption time was not a problem, because BdU signaled his U-boats several days before they reached their positions. This lead time was all the Allies needed for effective evasive action.[62] Shore-based HF/DF was never of comparable importance in this operational sphere, except in the absence of Ultra.

Compared to shipborne HF/DF, Ultra information had only a limited effect on the actual tactical operations against the U-boats in the Atlantic. But once there were sufficient escorts available, Ultra made it possible to send independent support groups, as well as VLR aircraft, to reinforce those convoys in real danger. This finally brought about the turning of the tide in the Allies' favor in May 1943. That summer, Ultra also allowed the U.S. Atlantic Fleet to send escort carriers and their hunter-killer groups to precisely the areas where the Germans planned their resupply operations. Then tactical seaborne HF/DF could assist by warning of U-boats in the immediate vicinity.

In spite of the widely understood technology that went into the Enigma machine, German mastery of electromechanical encryption led the Germans to overrate the security of their own device. In the similar case of high-frequency direction finding it led them to underrate the potential of the D/F equipment of their opponents. These mistakes were to prove fatal to the U-boat war.

CHAPTER 4

The Secret Weapon: Huff Duff

O n 31 May 1916, when the German High Seas Fleet sailed from Wilhelmshaven to the North Sea, the move was immediately detected by the British shore-based D/F network. This set the British Grand Fleet in motion, and the two forces met at the Battle of Jutland. Direction finding was pivotal in bringing about this long-expected yet indecisive engagement; so too was codebreaking. The Royal Navy's intelligence staff in Room 40 (where the codebreakers worked) had done a brilliant job of intercepting and deciphering German radio traffic, providing exact information about the movements of the enemy fleet. Because of lack of cooperation from the operations division, however, much of the information never reached commanders on the spot. This accounted, in part, for the inconclusive outcome of the battle.[1]

In spite of the failures at Jutland, direction finding and decryption had both proved their importance to naval warfare. The British had also used D/F to detect the presence of German submarines by their low-frequency radio transmissions, and both sides in World War I employed direction finding for navigation.

Direction-finding equipment seeks to determine the direction of arrival of a radio wave in order to establish the location of either the wave transmitter or the receiver. Thus a ship or an airplane can check its location by plotting the direction of arrival of signals from two or more known transmitting stations. The intersection of the lines of arrival provide a fix, marking the position of the receiver. Working the other way, the location of a vessel in distress can be calculated by triangulation when two or more receiving stations combine the bearings they obtain on the vessel's radio distress signal.[2]

Direction finding is a general term affected by a wide variety of phenomena and therefore encompassing many different types of devices. Direction finders may be intended for use in obtaining bearings on cooperative radio signals, or they may be designed to respond effectively to noncooperative signals. The technology may be implemented using receivers at land-based stations, but if receivers are to be placed on ships or in airplanes, the equipment must be quite different. Direction-finding requirements for handling low- or medium- or high-frequency radio waves ("long," "medium," or "shortwave") also vary considerably.

In the high-frequency range, direction finders may receive signals from sky waves that bounce off the ionosphere, from ground waves that travel close to the surface, from ground-reflected sky waves, or from multipath combinations of sky and ground waves. These need to be correctly identified if the high-frequency direction finder is to be used for long- or for short-distance detection, for general operational or for strictly tactical military purposes.

For example, shore stations generally take bearings off the sky wave, which enables them to locate transmitters several thousand miles away. When warships equipped with HF/DF are in an operational area, however, in addition to receiving sky waves they may also take bearings from ground waves, which do not travel more than twenty-five to thirty miles before dissipating. This means the ships are within striking distance of the transmitter.

In 1909 Oscar C. Roos of Cambridge, Massachusetts, applied to the U.S. Patent Office to register an "Apparatus for Determining the Direction of Space-Telegraph Signals." The patent (No. 984,108) was granted in 1911, and from such initial developments early in the twentieth century, direction finding went on to become an important navigational aid. It was also used to track the flight of meteorological balloons, to locate storm fronts, and to discover the location of illicit or enemy transmitters. The technology was employed in this latter way with some success during World War I. Despite the rudimentary nature of the equipment, it was capable of exposing the presence of German submarines when they sent out radio messages.[3]

Unlike sonar and radar, direction finders are passive listening devices, the airwave equivalent of the underwater hydrophone. When used for navigation, the task of position-fixing systems is relatively simple. Transmissions are beamed out from stations at known locations. The signals are steady, continuous, and on established frequencies. These cooperative navigational systems, such as the radio compass, are an important form of direction finding. Anyone can tune a receiver to these freely accessible transmissions and figure out his own location on a chart.[4]

Even in peacetime, though, a signal such as a distress call from a ship may be very brief. Radio direction-finding equipment initially required a number of adjustments and calculations in order to establish a bearing on a signal; this could take several minutes. A distress call may also be on a frequency that does not come in clearly at all receiving points, which further impedes the taking of a bearing.[5]

So even transmissions on which bearings are taken in many normal, civil applications can be noncooperative. Of course, when someone is deliberately seeking to hinder reception of his signals by unauthorized parties, he will certainly transmit on changing frequencies and in the briefest possible time.

Generally speaking, all radio direction finders have three major components: a directional antenna, or wave collector system; a radio receiver from which the directional information will be derived; and a display-and-indicator system that puts the information from the other two components into a form that can be used by the operator.

An ideal direction finder would be capable of covering a wide radio frequency band (the wider, the better the chance of finding the same frequency as the transmission), 360 degrees of azimuth (the horizontal angle of a bearing on the horizon, measured clockwise from the north), and 90 degrees of elevation. In other words, a direction finder should be able to detect transmissions on as many frequencies as possible, arriving from whatever direction and whatever elevation. It should also be direct reading (direction indicated simply on a dial or screen), and it should operate as fast as possible (the aim being an instantaneous response).

The direction finder needs to take into account and make adjustments for all sources of signal deviations. There are many of these, and they will be discussed below. The product of this "perfect" direction finder would be accurate and reliable bearings that were immediately available, even from very brief transmissions.

Initially, interest in radio direction finding focused on long waves, that is, on waves in the low-frequency range. This range is between 30 and 300 kilocycles, which is a wavelength range of 10,000 to 1,000 meters. Low-frequency (LF) transmission is good for navigation because it is dependable at all distances intermediate to its extreme range, and in all directions. LF waves follow the curvature of the earth's surface; reception is extremely reliable and not subject to the seasons or atmospheric conditions. Dependable communication can be obtained on low frequencies over distances up to one thousand miles by the ground wave alone. All direction finding during the First World War, whether it was from the chain of land-based receiving stations or from shipboard devices, took place in the LF range.[6]

Because of its usefulness for navigation and its potential military applications, direction finding was quickly adapted for deployment at sea. Shipboard direction finding on long waves was accomplished by means of a rotatable frame antenna. This antenna would be turned manually until a minimum signal was received, which was when the frame was at right angles to the oncoming wave beam. Then the bearing would be read off a 360-degree scale. During World War I some British and American battleships were equipped with early models of this kind of direction finder, for example the U.S. Navy's type SE74.

By 1918 the Radio Test Shop of the Washington Navy Yard had already designed the U.S. Navy's first aircraft direction finder, the model SE950. At the same time, the navy selected sites at the entrances to principal Atlantic seaports for installation of radio D/F stations. Three such stations were established to guide ships safely into New York harbor, and they began providing fixes for vessels in the vicinity on 26 December.[7]

Long-wave direction finding continued to be important in radio navigation, and shore stations were used to guide aircraft as well as ships. Long-wave, low-frequency direction finders could also be mounted in planes for navigation. Long-wave direction-finding equipment, however, could not receive the high frequencies on which the German U-boats of the 1930s began transmitting.

Between the wars there had occurred a great deal of research on shortwave, or high-frequency (HF) radio. Shortwaves range in length from 100 to 10 meters, and in frequency from 3 to 30 megacycles. In the early 1920s amateur radio operators using very moderate power discovered that their high-frequency transmissions not only extended to the horizon, but that they then frequently reappeared at distances far beyond.[8]

Long-distance transmission via sky waves is made possible by the existence of a series of ionized layers in the earth's upper atmosphere collectively forming the ionosphere. These layers act as a "mirror" to bounce the sky waves back to earth. High-frequency radio waves can cover long distances, but only when reflected from the ionosphere in single or multiple hop paths, and there is no reception at all in the skip distances between touchdowns. In general, reception off HF sky waves can be both elusive and tricky. It is subject to fading and to erratic changes with the seasons; changes between night and day; variations in the number, density, and height of the ionized layers; and atmospheric conditions.[9]

All these factors contribute to considerable variation in the returning sky wave and to fluctuating length of hops. The well-known "night error" condition produced by ionospheric reflection, to which HF direction finding is especially susceptible, increases the likelihood of false bearings. Yet HF transmission is also capable of providing around-the-world communication. Once that had been established, it quickly became irreplaceable.

Radio waves launched from an antenna travel outward in all directions. They are able to penetrate nonmetallic objects but are stopped dead by metals or fine-meshed wire screens. They seem to travel best in free space, and they do not need a medium of propagation as do water waves.

Beyond these generalizations, however, there is little that can be said about the propagation of radio waves without specifying wavelength or frequency. The behavior of the different waves, and therefore their usefulness, differs considerably. In general, as the frequency goes up the radiations tend to travel in more nearly straight lines, like light waves. So unlike LF waves, HF waves do not follow the curve of the earth. Also, like light waves, HF waves are relatively easily reflected and refracted, which makes them much less reliable than LF waves and therefore much less useful for navigation.[10]

An antenna designed to intercept HF waves is accessible to the three kinds of signals: the surface or ground wave, the sky wave, and the sky wave reflected from the ground. The ground wave will give a consistently accurate bearing if there are no geographic or other obstructions in the wave path. Low-angle sky (ionospheric) waves, or rays, may produce fairly accurate bearings. But high-angle sky waves are subject to the serious deviations already described. Ground reflected sky waves are even more unreliable, variable, and often produce no bearing at

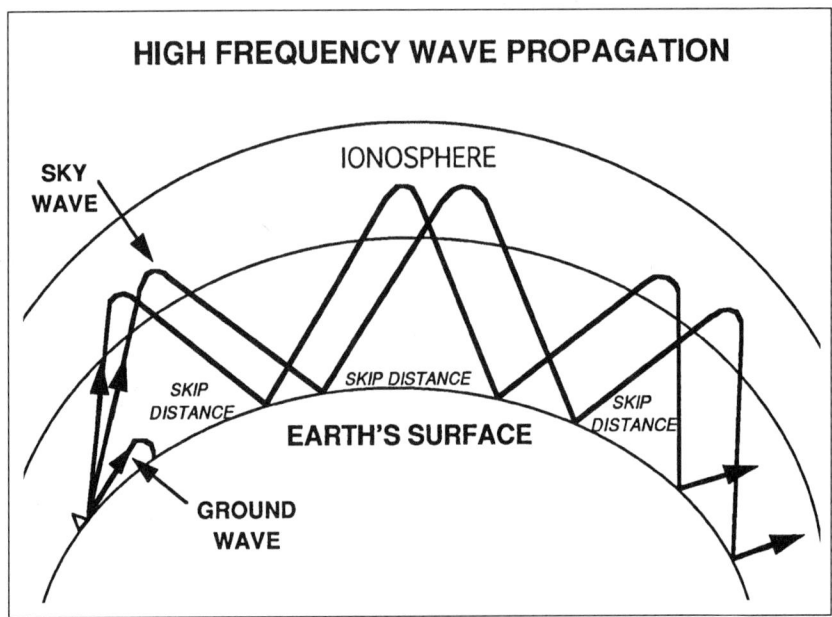

HIGH FREQUENCY WAVE PROPAGATION

SKY WAVE

IONOSPHERE

SKIP DISTANCE

SKIP DISTANCE

SKIP DISTANCE

EARTH'S SURFACE

GROUND WAVE

Fig. 4.1. High-frequency wave propagation chart.
Drawing by Brooke C. and Alexandra T. Williams

all. Frequently the receiver even picks up multipath signals: sky and ground waves combined.

The difficulty of distinguishing which kind of wave is being received requires more precision in the instruments than does low- or medium-frequency direction finding, as well as considerably more skill and experience on the part of the operator. But for military purposes, and especially if HF/DF is to be used tactically, it is essential to make such distinctions. A transmission received on a sky wave could have been made from thousands of miles away. A transmission received off a ground wave, however, can have traveled no more than thirty or so miles. In World War II, receipt of a ground-wave signal therefore meant that a transmitting enemy submarine was within close proximity, most likely within visual contact range of the receiver.

Like reading sonar and radar returns, effective HF direction finding, especially on noncooperative signals, is partly an art. The skill of the operator, especially with early models, was at least as important as the equipment itself. The sky wave is at best unpredictable, and the HF ground wave is rapidly attenuated with distance. Moreover, the difference between the two is not at all obvious to the inexperienced HF/DF operator, especially when working at top speed on a short, hard-to-tune signal.

After World War I, research on direction finders mostly concentrated on their use for the location of thunderstorms. In the 1920s and 1930s, major progress was made in high-frequency direction finding. Laboratories and workshops in several different European countries, as well as in Japan and the United States, were engaged in research and development. Each country was producing similar and steadily more sophisticated equipment, aimed at providing increasingly accurate and reliable bearings.[11]

Among the scientists working on direction finding, two Europeans were already developing devices that were to have significant military applications in addition to their meteorological and navigational uses. Dr. Robert Watson-Watt, the Scotsman often referred to as the father of radar, also described himself as "the proud parent" of cathode-ray direction finding. The cathode-ray tube indicated the received transmission on an easy-to-see screen instead of the usual dial. As early as June 1923, while at the Radio Research Board Station, Aldershot, England, Watson-Watt had a cathode-ray direction finder (not HF, though) in operation. The "kathode ray oscillograph" used in this device, however, had been developed by an American, O. Webb, who had been demonstrating it to the Institution of Electrical Engineers in London. Watson-Watt prevailed on Mr. Webb, who only had two, to give him one tube.[12]

Continuing his work in the field of radio D/F (which included both radar and D/F), Watson-Watt published an article in 1926 on an instantaneous direct-reading radio goniometer. His cathode-ray direction finder, however, had not yet been applied to high frequencies. That same year a young French engineer, Dr. Henri Busignies, obtained a French patent on his own practical model high-frequency direction finder.[13]

Similar work was going on in the United States, some of it for the U.S. Navy at the Naval Research Laboratory, but much of it also in corporate research labs. In 1937 two scientists working for Bell Telephone were granted a patent on an "Electrical Indicating or Measuring System" for production or control of radio waves. That same year Henri Busignies, through the International Standard Electric Corporation of New York, applied for a patent on a "Radio Direction Finder." In 1938

Donald S. Bond of the Radio Corporation of America submitted a design to the U.S. Patent Office for his "Ultra High Frequency Radio Direction Finder," and in 1940 Josef Plebanski filed a patent application in the name of the Radio Patents Corporation of New York for a "Radio Direction Finding System."[14]

Most of the experiments conducted in the interwar years used an HF direction finder based on a fixed antenna system. As with radar, much early research in direction finding concentrated on the most appropriate antenna configuration for the particular task at hand. The antenna array of a direction finder originally patented in 1919 and named for its British inventor, Frank Adcock, eventually became the standard long-range, land-based high-frequency D/F antenna in both Britain and the United States. Adcock had been a signals officer in World War I and then worked as a research chemist at the National Physical Laboratory where Watson-Watt also spent a number of years.[15]

The U Adcock antenna consisted of four identical vertical poles, or elements, at the corners of a square. The square was usually oriented with one diagonal running north-south and the other east-west. The signals that the antenna grasped from the air were brought by buried coaxial cable feeders to a central receiving station three to four hundred feet from the antenna. The station housed the other components of the high-frequency D/F, the receiver and the display system giving a relative bearing in degrees, as well as transmission facilities to forward data to a central processing office.[16]

Besides producing significant developments in antenna systems, research in high frequencies in the 1920s and 1930s also affected changes in the other components of direction finding. There was steady development in receivers and goniometers and in the display or indicator systems. Efforts were also made to improve the reliability of reception by better understanding the causes of signal deviations.

By World War II most radio transmission was by continuous wave, that is, by radiotelegraphy. Continuous radio waves were received through the feeder cables from the antenna, and the receiver converted them into useful form. This useful form consisted of either an audio signal (through earphones) or a visual signal that was displayed on a scale or indicated by a dial or, best of all for speed and accuracy of response (as both Busignies and Watson-Watt understood), on a cathode-ray screen.

A direction finder differs in function from a radio receiver in that it

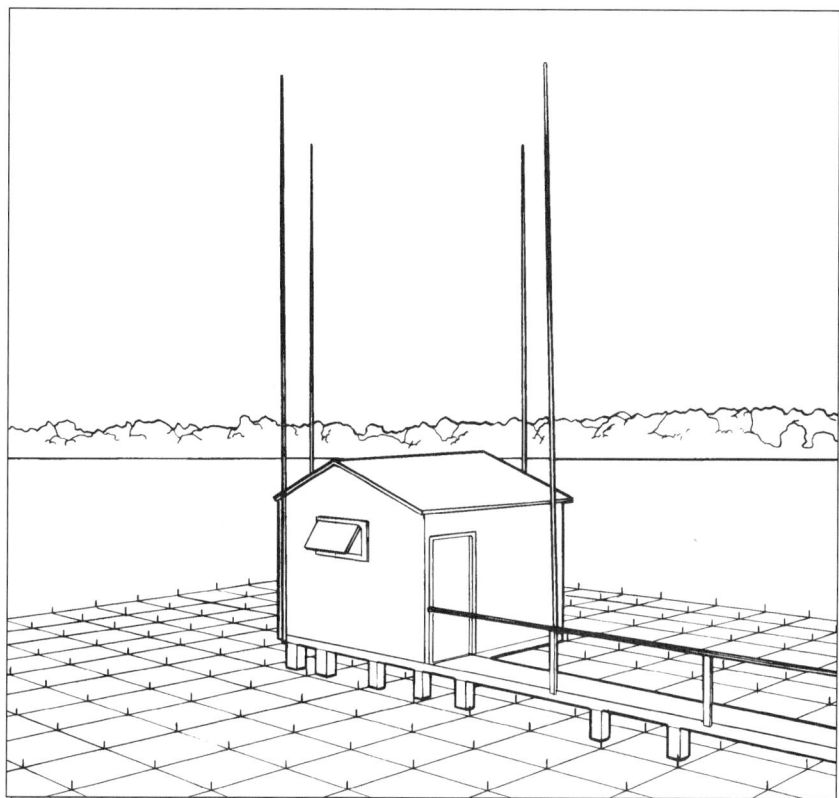

Fig. 4.2. Land-based HF/DF Adcock antenna shack.
Drawing by Roger G. B. Broome

does not just receive the radio waves, it also needs to be able to ascertain the bearing (direction of arrival) of those waves. Therefore, in addition to a standard radio receiver, direction finders also need a radiogoniometer. A goniometer is an instrument consisting of a transformer circuit that extracts directional information from the received radio waves.[17]

Most direction finders operate by taking bearings on signal minima, or null, as it is generally easier to tune to the minimum signal strength than to the maximum. Pre–World War II goniometers, therefore, generally had a pointer for taking bearings on the signal minimum, which was then indicated on a scale. On most of the existing D/Fs, bearings were taken by manually adjusting the goniometer to the signal minimum and reading the pointer's position off the scale.[18]

At this point only the "sense" of the incoming signal remained to be determined. When an indicator shows a line of bearing, it registers a line across a 360-degree scale. It does not tell from which of the two possible directions the signal is coming. Initially sense determination was largely a function of the operator and had a great deal to do with his experience. A distress signal from a ship, for example, would not be coming from a large landmass. The only other way to obtain a sense determination at sea was to steam along the line of bearing and check for an increase or decrease in the signal strength.[19]

Gradually more sophisticated equipment was developed to indicate direction as well as bearing. This usually involved turning a switch or pressing a sense button, which would turn a pointer to the observed bearing and then its reciprocal, noting the change in signal strength. If this was done correctly the pointer indicated the correct sense, but sense ambiguity remained a problem in direction finding.

It was a particularly serious problem at sea, where logic usually cannot help the operator. Securing proper sense operations over the wide frequency range required to track U-boats in World War II was a major difficulty, which had by no means been solved by 1939. It remained a serious obstacle to the British navy's efforts at shipborne HF/DF until early 1941, when a Polish engineer, W. Struszynski, working in England, developed a sense antenna.[20]

Direction finding prior to World War II was a complicated procedure subject to many variables of equipment and natural environment. Even with well-trained operators, the results of early direction finding were generally imprecise except at medium frequencies, which were much more reliable. The introduction of HF radio transmissions had greatly magnified the difficulties of direction finding. The ionosphere was an irregular and changeable reflector, and there was no reception at all in the skip distances between surface rebounds. Moreover, high-frequency transmission techniques had been developed so that signals could be extremely brief; it was especially challenging to tune to them fast enough to obtain an accurate bearing.[21]

In 1936, when Britain almost went to war with Italy over Abyssinia, it already had five radio-interception or Y stations in operation—two in the United Kingdom, one on Malta, and two in the Far East. The United States too had a number of radio-intercept stations by this time, although they were called receiving stations. Each British station was part of an integrated operation that eventually collected, sorted, and sent on to operational headquarters both D/F and encrypted message information from the whole network.

By the summer of 1937, when the Operational Intelligence Centre for the Royal Navy was established, bearings from these naval D/F stations were first worked up into a tentative D/F plot. The Spanish Civil War afforded some more practice in plotting, and Italian submarines were tracked as they maintained a blockade for General Franco. Nevertheless, it was July 1939 before a twenty-four-hour watch system was set up in Britain to monitor the information from proliferating D/F stations.[22]

By the outbreak of World War II, because of unwillingness to spend on defense, there were still only six high-frequency and four medium-frequency D/F stations in Britain, three in the Mediterranean, and two in the Far East. Eventually the coverage was extended to include eight more in Scotland and England, as well as stations in Gibraltar, Iceland, and the Azores. What this provided was a long-range shore-based system for general surveillance over wide areas.[23]

In addition to errors arising from propagation effects, high-frequency direction finding may also be subject to small errors caused by local site irregularities, as well as to larger errors due to reradiation from objects within a radius of up to a mile. The selection of sites for land-based direction finding stations is therefore of great importance. The best site is on very level land with uniform ground conductivity. But even when the site is chosen with the utmost care, errors often remain.[24]

During World War II, the enormous expansion in direction finding created a network of land stations ringing the coasts not only of Britain but of the United States as well. The demands of this network meant that some direction finders had to be erected on sites that were judged to be poor in terms of conductivity and freedom from obstacles.

But the important role expected for direction finding, which mandated the use of even imperfect sites, also gave a stimulus to research. In the course of the war the various sources of site error as well as of instrumental error were gradually identified, and remedies were found. Earth mats were designed to deal with sites of poor or irregular conductivity. Other consistent site errors were dealt with by a correction table obtained by calibration with a test oscillator. The oscillator was used to produce a high-frequency radio wave, and the standard deviations at each site could be plotted and allowed for. Calibration became a significant part of the effective operation of high-frequency direction finders.[25]

While such developments resulted in a steady improvement in the accuracy of bearings, a shortage of qualified operators was initially an even more severe handicap than the limitations of the equipment. Comdr. Patrick Beesly, RN, an intelligence officer who served in the British Admiralty's Submarine Tracking Room during the war, would have agreed with Dönitz's early assessment that the danger to U-boats from direction finders was not very great. Beesly was very much aware of the weaknesses, at least in the British shore-based HF/DF network.

The loss of France meant a loss of cooperating shore stations, which took time to replace. Because the more bearings obtained on a transmission, the more precise would be the "cut," or fix, it was important to have as many receivers in operation as possible. It took bearings from six or more different shore stations to establish a U-boat's position to within twenty-five miles or so, and this still assumed a margin of error of around three degrees.[26]

Nevertheless, the rerouting of convoys based on information from HF/DF shore stations began in 1940 after an observation vessel had established German procedures. But the general inefficacy of the system in those early days was demonstrated when rerouted convoys not infrequently ran right into German weather-reporting ships. The ships' positions should have been clear to HF/DF, as they had been transmitting regularly from the same location for some time. It appears that Dönitz noticed the inadequacy of early HF/DF suggested by these incidents, and his wariness of the device relaxed. It is significant that he did not pick up on later suggestions of the possibility of an improved device.[27]

In the United States, meanwhile, development of a shore-based direction-finder network comparable to that of Britain was also being set up in the late 1930s. The States too was building on World War I experience—direction finders from the U.S. Naval Base at Brest in the Bay of Biscay had been used to locate German submarines. American attention was initially focused on the West Coast because of a growing rivalry with Japan. By 1939 Japanese warships and merchant vessels using short-wave communications were being tracked in the Pacific by American HF/DF stations.[28]

But there was an even greater shortage of qualified operators in the

U.S. Navy than there had been in the Royal Navy; in the beginning, the accuracy of bearings that the Americans obtained did not approach that of the more experienced British. One of the difficulties of coalition warfare is setting up systems that institutionalize a willingness to learn from each other. Scientific exchanges between the future Allies ultimately were very influential in improving the technologies of both Britain and the United States. But this cooperation, which had been virtually abandoned in the interwar years, got off to a slow start, and it was to be a while before any organized, structured mechanism for regular consultation was set up.

In the fall of 1940 Churchill sent Dr. Henry Tizard to the United States with the cavity magnetron that had been developed in Britain and that made centimetric radar possible. This "gift," proffered to coax support from the States in Britain's most desperate hour, opened the way for negotiations that ultimately led to full technical cooperation between the two countries. The cooperation was accomplished through the American London Mission and the British Admiralty Delegation to Washington.[29]

In the winter of 1940–41 the British were still mostly reluctant to share their scientific information with the United States, even though President Roosevelt had decided that Britain was to be given access to all American technology except for the coveted Norden bombsight. However, after the Pearl Harbor disaster the United States requested and obtained the services of Dr. Watson-Watt to upgrade the American radar-defense network. By this time Watson-Watt had moved on from storm location by direction finding to setting up air-defense radar systems in his position as scientific adviser on telecommunications to the Air Ministry.[30]

Help with the American HF/DF network appeared in a different way. Fortuitously, on the last day of 1940 a team of French scientists from ITT's Paris office arrived in New York. One of the team, Henri Busignies, was an expert in high-frequency direction finding. He had worked on systems for the French navy and air force. Under his guidance a prototype of the French instantaneous automatic-indicating HF/DF was reconstructed; it was more accurate and required less skill to operate than anything then available in the United States. It was also more advanced than the British FH3 type, which was neither instantaneous nor direct indicating. The U.S. Navy adopted the French device and installed it at all new shore stations, such as Casco Bay, Maine.

Between 1939 and 1941, radar and other electronics research was being undertaken in Canada (and other Commonwealth countries), as well as in Britain and the United States. But though Canadians developed, among other things, a way of recording HF/DF bearings semi-automatically, their escort groups usually had less technical equipment (often of inferior quality), and their work in this field has remained little known. What does seem clear is that initially there had been substantial duplication of effort among all three countries.[31]

After the U.S. shore-based D/F network was established, the Allies soon evolved a system for cooperative sharing of direction-finder bearings, though clearly the British contribution was initially more valuable. The development of new equipment in the United States, as well as an improved training program for operators, soon helped to even out the contributions. Eventually, fifty-one radio-intercept stations in Britain and North America could triangulate almost any North Atlantic broadcast and send the information to centers in London, Ottawa, or Washington.[32]

The fitting of HF/DF in ships was found to be much more difficult than producing adequate land-based equipment. MF/DF equipment (used to detect the medium-frequency homing signals used by U-boats) was easily installed and gave accurate bearings when tested on vessels at sea, and by 1939 it was already available in many British ships.[33] But research conducted by the National Physical Laboratory in England before the war showed that shipboard direction finding was "wildly inaccurate" at high frequencies, particularly those higher frequencies being used deliberately by the Germans.[34]

It was not understood why this was so, nor what could be done about it. The German high-frequency transmissions could be clearly heard but could not be D/F'd, which meant that they could not be made to yield credible bearings. The apparatus seemed to indicate a direction almost arbitrarily.

The effective tactical use of high-frequency direction finders by escort vessels had to await the resolution of the problem. This required the development of suitable antennas and, equally important, their correct placement in the vessel. Once these practical problems were solved and an effective shipboard antenna was linked to a display system with sufficient speed of reaction to capture Dönitz's brief high-frequency transmissions, then it finally became possible to regularly obtain readable bearings at sea.

When dealing only with ionospheric reflections the U Adcock remained the preferred antenna down to ranges of about two hundred miles. During World War II the Allied naval shore stations all operated with Adcock antenna systems. For high-frequency direction finding at distances below two hundred miles, however, and for ground-wave reception, a quadrature loop array was found to have a somewhat more accurate performance. Also, the Adcock antenna, with its system of four widely separated elements grouped in a square, was obviously impossible to install on ships or planes.[35]

Ultimately the problems of designing suitable direction finders for short-range, mobile, shipborne use led to the development of the Bellini-Tosi loop system. Even the Bellini-Tosi antennas were rather large and unwieldy, initially, but the windage encountered on the upper deck of ships mandated and led to a gradual reduction in size. Concurrent developments in amplifiers from the early 1920s, among other things, had made such reductions possible. The trend toward miniaturization is particularly clearly demonstrated in the antenna field. But the problem of stress on the relatively large antennas required by high-frequency signals was never overcome for aircraft, and there was no airborne HF/DF.[36]

The choice of sites for antennas on ships was as important as it was on shore, and a similar concern for site errors had to be addressed. Many of the same general common-sense principles prevailed in each case, and yet the British had not successfully resolved the issue on the outbreak of war. Theoretically direction finders could be made to work accurately at sea on high frequencies, but how?

Britain had produced a high-frequency D/F set for ships before the war started, but its performance in trials was "very inferior." In large part due to the "numerous technical shortcomings of the apparatus" itself, the results were even more disappointing than had been the initial results from shore-based equipment.[37] The Admiralty Signal Establishment (or Signal School) meanwhile developed another HF/DF device on the old method of aural reception. This device, the FH3, was first installed in the destroyers *Gurkha* and *Lance* in July 1941. The HF/DF antenna (the Struszynski model) was placed at the top of the foremast in place of the radar antenna. But the commander in chief, Western Approaches, insisted that all his ships have radar, so the next escorts to get HF/DF had to place the antenna on a special pole mast from where the reception was not nearly so good.[38]

At this point radar appeared to be a much more promising technology, and it was certainly more glamorous than HF/DF, being active instead of passive. For a number of years it continued to be thought that the two devices could not be used to best effect on the same vessel, as they each needed to be placed atop the highest mast. Also, radar transmission interferes with HF/DF reception and cannot take place at the same time. Clearly competition for scarce developmental and manufacturing resources, as well as for space on crowded upper decks, hindered the production of an effective shipborne high-frequency direction finder by the British.[39]

The ineffectiveness of British HF/DF-equipped escorts early in the war, and their lack of success against the U-boats, had evidently made an impression on someone in the U.S. Navy. A 1942 memorandum in the files of the Tenth Fleet noted that by November 1941, the aural HF/DF device had been somewhat improved. But still the difficulty of interpreting ambiguous readings was such that the Royal Navy "decided to abandon the fitting [of HF/DF] in favor of certain types of radar which require the same masthead site as the HF/DF aerial system." The memorandum continued that even some of those HF/DF sets that had already been fitted on Royal Navy vessels were removed in favor of radar.[40]

Meanwhile, the Signal School was working on a new device. All the basic principles and the circuitry for this new apparatus had already been worked out long since by Watson-Watt and his team at the Radio Research Station. It was to be based on the use of a cathode-ray screen for visual presentation, and it incorporated the improved sense-finding capability that was so important to shipboard direction finding. Sense ambiguity could be a crippling handicap if, as often happened, a single HF/DF-equipped vessel had to run down the one bearing it had itself obtained, with no possibility of triangulation. In a close parallel to the American experience with HF/DF, the British navy was also able to turn to an experienced civilian manufacturer, the Plessey Company, for supply of the new equipment.[41]

In October 1941 the first experimental cathode-ray HF/DF—the FH4—was installed on the ex-U.S. Coast Guard cutter *Culver*. But in January 1942, when the *Culver* was lost with its HF/DF aboard, this somewhat retarded the whole program. In general, fitting of the device did not get under way until May 1943.[42]

The U.S. Navy's early experience with the development of shipborne HF/DF was similar to that of the Royal Navy. Although ITT engineers had a device ready in 1941, little was done to put it into operation. An NRL report of June 1942 gives one reason why:

> It is well known that naval shipboard direction finder performance has been much less satisfactory at high frequencies (1-30 Mc) than at lower frequencies (below 1 Mc). From a navigation viewpoint, all known types of high frequency direction finders, including Adcock as well as loop type direction finders, have given such poor results, particularly on long distance (i.e. skywave, abnormally polarized) transmissions, that high frequency shipboard direction finder developments were effectively discouraged for a considerable time.[43]

This report was written by the NRL's D/F expert, Dr. Maxwell K. Goldstein. While the British continued their search for a practical way to get more effective HF/DF bearings at sea, the same problem was being addressed in the United States. In 1942 Dr. A. Hoyt Taylor, civilian head of the NRL, had handed the challenge to the dozen or so professionals and technicians of the Radio Direction Finder section. The wartime head of the section, Max Goldstein, was a short, intense, inventive scientist with a flair for finding practical solutions to highly complex technical issues. Goldstein also found a particularly striking way of describing the problem with shipboard direction finding in the early war years. He often said that at high frequencies, "Detecting the direction from which the transmission arrived was as baffling as standing in the hall of mirrors in a fun house, seeing yourself reflected a hundred times and trying to discover which reflection was true, and which was a reflection of a reflection of a reflection. For this was how the incoming signal appeared—echoing endlessly from all directions. Why?"[44]

Just as it was important for land-based direction finders to have a clear ground site with no obstacles capable of reradiation within several hundred yards, so too the shipboard antenna should be positioned to have the clearest possible field of reception. Obviously, the greatest freedom from errors and problems could be obtained by placing the antenna high above the ship's own structure. But mast height on a ship is always restricted by considerations of stability as well as by the need to clear a decent field for antiaircraft fire.

A tall-mast solution could not always be implemented, as the British found, nor, as Dr. Goldstein discovered, did it completely resolve the ship's site errors, which generally proved much more formidable and harder to overcome than those on land. For one thing, a multitude

Dr. Maxwell K. Goldstein of the NRL. *Courtesy of Barbara Goldstein Koz Paley*

of electronic devices on board ship (radio and various kinds of radar among them) competed for antenna space in a limited area. And there were other potential sources of deviation errors to be considered—in the metal hull, the masts and struts, the stays and other rigging, the fittings, the funnels, the superstructure, and the ordnance. Any of these might form a competitive loop that could throw off the accuracy of the bearing. Nor was it possible, except in very large vessels such as carriers, to use the ship's main transmitter at the same time as taking D/F bearings, because a certain minimal distance had to be maintained between transmitter and receiving antennas.[45]

After intense investigation, Goldstein and his team concluded that the ship itself was the problem. The "tangle of metallic clutter above-deck" received the incoming signal just as the direction finder did, except that each element also acted as an antenna, emitting a signal of its own when "excited" by the arriving radio wave. It was these signals from the ship's own fittings that were being received from all directions by the direction finder. The result was Goldstein's electronic "hall of mirrors," in which the D/F could not separate the incoming signal from its various echoes.[46]

There was no simple, practical solution to this problem; rather it required a detailed and carefully conducted procedure. Once a direction finder was installed on a vessel, as many as possible of the variable errors were dealt with by elimination or modification. Even hemp ropes, harmless when dry, can become partial conductors when wet. The closed loops they could form needed to be broken up by insulators.

The remaining untreatable sources of deviation were handled, as on shore, by a careful calibration process. The constant errors were determined by comparison of the difference between observed and true bearings at as many points as possible in each quadrant. Then the proper corrections could be applied when the direction finder was in operational use. And because the deviation errors were not uniform but varied from position to position, compensation had to be calculated; it could not be accomplished by an adjustment of the equipment.[47]

If a bathroom scale gives uniform readings of five extra pounds, it can be adjusted downward by 5 pounds. But if it is 5 pounds off when a person of 120 pounds is weighed and 12 pounds off when a person of 180 pounds is weighed, a separate calculation needs to be made for each different weight. So it was with the calibration of direction finders. Site errors for land-based direction finders were relatively controllable and rather small, so that very precise equipment made a significant difference. The larger errors unavoidable on steel ships, overloaded with competing electronic devices, meant that a larger margin of error had to be tolerated.

And still, this improved understanding of deviation errors by no means resolved the challenges of placing high-frequency direction finders at sea. Just imagine trying to tune a receiver and obtain a bearing (after consulting the calibration chart) from an uncooperative signal lasting only seconds, while tossing and pitching in a cold shack on the deck of a destroyer in the North Atlantic.

This helps to explain why German U-boat Command felt secure in the use of high-frequency transmissions for communication in wartime. Even if their signals could be detected, it was believed that they were too brief to give an accurate bearing, that the antennas required to receive them were too large to be mounted on ships, and that the signals would be received by shore stations too far distant to be of tactical or perhaps even operational use. By the time a convoy escort could be directed to the general area, the U-boat would be long gone.

This was a rational picture of HF/DF capabilities in 1939. Even as late as August 1942, it took anywhere from two hours and twenty-four minutes to eleven hours and fifty-seven minutes for U-boat reports (from whatever source) to reach American convoy escorts through the office of the commander in chief ashore. This was too slow for most tactical applications.[48]

With D/F aboard, on the other hand, escorts could act directly against the U-boats as soon as they intercepted their high-frequency transmissions, and without the inevitable delays incurred by waiting for notification of fixes from shore stations. In spite of the clear advantages of shipborne HF/DF, however, it was not deployed as early as it might have been. Delay in its introduction gave credence to Dönitz's belief that shipborne HF/DF was impossible.

What eventually proved Dönitz wrong was the development, during the course of the war, of new, reliable, shipborne equipment, designed specifically for noncooperative direction finding on transmissions of very short duration. The success of this device could negate the U-boat's radio communication advantages of brevity of signal and mobility of transmitter. High-frequency direction finding evolved as a part of the whole spectrum of electronic countermeasures, and it played a crucial role in the development of twentieth-century electronic warfare.[49]

Just as there was a slow evolution in design and effectiveness of HF/DF equipment, so too appropriate ASW tactics had to be developed for its use. By the end of January 1942, twenty-five of the larger British escorts and rescue ships had been fitted with the old aural FH3 sets. Partly because of the loss of the experimental FH4 on the *Culver* earlier in the month, though, it was not until March that the first FH4 saw action on the destroyer *Leamington,* escorting a convoy to Madagascar.[50]

By August there were about seventy vessels in the Royal Navy fitted with HF/DF, the majority the old FH3s, and although there were to be thirty FH4 sets available by the end of the year, delivery of more than that was not to start until May 1943.[51] So it was mostly the old aural FH3, with all its quirks and limitations, that played such an important part for Britain in the convoy battles of 1941, 1942, and 1943.

Because HF/DF had a ground-wave range of up to thirty miles, far exceeding radar and sonar, it was shipborne HF/DF that often gave the first indication of U-boats in the vicinity of a convoy. And because Dönitz's U-boats followed a predictable pattern of transmissions once in contact with a convoy, it was possible to form a fairly accurate estimate of what they were up to and to plan evasive or offensive actions accordingly.

Each U-boat message had required prefixes indicating the priority of the message. The length of the message, too, was indicative of its content. The first patrolling U-boat to sight a convoy would make a brief high-priority report, giving only the convoy's position and heading. Subsequent amplifying reports would be of lower priority and longer, and they included information on the size of the convoy and on the number and type of escorts. For obvious reasons it was particularly important to indicate the presence of an escort carrier with supporting aircraft. The requirement by BdU for the shadowers to continue calling in periodically made them still more vulnerable to detection.[52]

As soon as a U-boat's first signal of contact with a convoy was D/F'd, an escort was sent after it, down the line of bearing. Meanwhile the convoy commodore would order a change of course that would take the convoy out of range before the U-boat could surface again. This close-in defensive work done by shipborne HF/DF thus held off many attacks even after U-boats had found the convoys.

By late 1942 there were many reports of HF/DF-aided successes in evading U-boat attacks on convoys, and in some cases information supplied by HF/DF enabled escorts to carry out successful offensive attacks. Along with centimetric radar, shipborne HF/DF eventually made wolfpack surface attacks on convoys too dangerous to attempt.

This record impressed Rear Adm. R. M. Brainard, the commander of American Task Force 24, in Argentia, Newfoundland. On 27 October 1942, Admiral Brainard wrote to the commander in chief, U.S. Atlantic Fleet: "The HF/DF in British escort vessels has proved itself repeatedly to be a tactical weapon of importance approaching that of radar." But he went on to caution that "the ultimate effectiveness and efficiency of this new installation can only be attained when manned and operated by specially trained personnel. HF/DF cannot be manned, as can radar, with any degree of efficiency or satisfaction by men other than trained radiomen."[53]

Thus another reason for promoting the development of radar over HF/DF was that it required less training for its operation. Because of shortages of both equipment and trained operators, it was not until late 1942 that Huff Duff really began to play the important part of which it was capable in the war against U-boats in the Atlantic. And even then, the provision of properly qualified technical personnel continued to be a problem, especially for the Canadians.

The Royal Canadian Navy had expanded extremely rapidly from an insignificant coastal force in 1939, and it was persistently hampered by problems of inadequate training. This was especially true in the technical areas, where inexperienced and poorly trained operators of radio electronic devices had to deal with poorly maintained equipment. The Canadians were probably right to complain that they could perform better if they were not treated like stepchildren when it came to the allocation of new vessels and equipment. They had also lost some of their best young scientific graduates to British ships.

But some of the problems were of their own making. The Naval Service Headquarters in Ottawa, preoccupied with construction schedules, had been hesitant about modernizing to keep up with new technical developments. Canadian corvettes suffered from a shortage of gyro compasses and had primitive asdic sets. The service headquarters had also been slow to accept non-Canadian radar devices, and they resisted shipboard HF/DF until it had proved its efficacy. HMCS *Restigouche* was the first Canadian ship to have HF/DF installed in 1942, only because its commander scrounged a set for himself when he was refitting in Londonderry.[54]

At any rate, Admiral Brainard's words quickly became prophetic. Between 30 October and 5 November 1942, the forty-two-ship convoy SC 107 set off eastbound across the North Atlantic accompanied by Canadian Ocean Escort Group C4. The convoy was first set upon off Newfoundland, and ultimately sixteen U-boats joined the attack. Of the escort, only the *Restigouche* and the rescue ship *Stockport* were equipped with HF/DF. Although the HF/DF on the *Restigouche* was used to good effect by vigorous pursuit of bearings that held off numerous attacks, the convoy eventually lost fifteen ships, including one tanker and the commodore's flagship. The Canadian escort group was sharply criticized for a poor showing in its area of control—for lack of training, for poor coordination with the RCAF, and for inadequate maintenance and use of technical equipment, especially the radar, much of which had been inoperable.[55]

What the unfortunate experience of SC 107 so vividly demonstrated was that neither HF/DF nor any other device was a panacea. Success against the U-boats depended on reliable equipment in the hands of well-trained operators working in conjunction with every possible weapon, device, vessel, and aircraft that could be brought to bear.

The Royal Navy and the U.S. Navy had little success with HF/DF afloat until the rudimentary equipment generally available in 1939 had been replaced by the improved devices with properly installed antennas, which did not come into widespread use until 1942 and 1943. Until that time HF/DF had been developed only to the point where it could be used effectively ashore.[56]

Given the obvious tactical potential of shipboard HF/DF to precisely locate U-boats on the point of attacking convoys, it is surprising it took the Allies so long to produce, install, and operate adequate numbers of the device. This is one of the puzzles of the Atlantic campaign; another is the German failure to anticipate such a development. The technology had been available to the Allies from early on, but the lure of sonar and then of radar, had been too compelling. As a result, for almost the first three years of the war, the German high-speed, high-frequency radio transmissions had posed little direct danger to the U-boats.

Under the pressure of war, the allocation of resources for research and development and for training involves choices that are not always easy to explain later. Although direction finding was a familiar concept in 1939, there was still much that needed to be understood about high-frequency transmission and reception before it could be properly harnessed for tactical use at sea. Speaking for the U.S. Navy, Admiral King later admitted that "the development and application of electronics to warfare had scarcely commenced when we entered the war."[57] While this does not, on its own, explain the paradox of the availability of high-frequency direction finding and the slowness with which it was put to offensive use, it must be considered as a factor that retarded HF/DF's development.

Direction-finder plotting always remained something of an art. The vagaries of high-frequency reception meant that deduction based on experience was the most reliable way to coax reasonably accurate information from merely suggestive data. Jumping to the wrong conclusions could be dangerous, which was why it was crucial to have highly trained operators as well as superior equipment.

Once shipboard HF/DF came on line, however, it quickly affected the nature of the struggle against the U-boats. In 1942, during the time of the Ultra blackout from 1 February to 15 December, shipborne HF/DF began to play an important role. By this time many rescue ships and some escort vessels were equipped with the technology. At first these were British sets, mostly FH3s, and then, toward the end of this period, increasing numbers of American DAQs became available. As the case of convoy SC 107 demonstrated, though, it took training, coordination, and sufficient numbers of well-equipped escorts, as well as air support, before a really impenetrable screen could be maintained between attacking U-boats and their merchant-ship prey. Obtaining kills was even more difficult.

Not surprisingly, it was some months before Allied forces, even those afloat, grasped the significance of shipborne HF/DF as a new source of U-boat location information. In August 1942 a "Memorandum on the Employment of High Frequency Direction Finding Equipment in Ships" still noted: "It has been found necessary to train a certain number of officers . . . in order that they may . . . advise escort commanders of the value of DF bearings."[58]

The next few months, from November 1942 to March 1943, witnessed the greatest concentration of wolf packs against convoys. While the fiercest convoy battles of the war were raging in the North Atlantic, either British or American shipboard HF/DF was installed on increasing numbers of vessels. By then it had become such an indispensable apparatus that the earlier trend was reversed, and in certain ships air-warning radar sets were removed to make room for HF/DF. Finally, a lattice foremast was designed so that HF/DF and radar could both be accommodated efficiently.[59]

The few weeks between March and May 1943 were crucial for the anti-U-boat campaign. The Allied successes of those weeks were both fueled and later sustained by an organizational change in the prosecution of the Battle of the Atlantic. The Atlantic Convoy Conference signaled the first really effective effort to use the combined strengths of Britain, Canada, and the United States to maximum effect.

On 1 March 1943, British and American Combined Staff Planners issued a report that indicates the extent of the confusion of effort that had previously hampered the development of scientific and technological solutions to the U-boat war. The report, entitled "Measures for Combatting the Submarine Menace," makes a forceful case for joint planning: "In technical research, and the development and production of anti-submarine technical devices and weapons, there has been substantial duplication of effort between the Allied Nations. Advanced and satisfactory equipment, or tactics, developed by one of the Allies, has not always been available to and adopted by the others with the expedition required for most effective combined effort."[60]

Diverting convoys away from concentrations of enemy submarines, whether by Ultra or HF/DF information, had not been sufficient to keep the sea lanes open. What was needed was a way of hitting back at the U-boats, of attacking them actively, primarily to drive them away from convoys and additionally, if possible, to reduce their numbers by sinking them. For real-time U-boat location the escorts had to have direction finders on board, and they also had to have aggressive tactical procedures.

These procedures eventually depended on the availability of tactical air support launched from escort carriers, as well as on sufficient numbers of escorts to be able to detach some of them to actively hunt down HF/DF-located U-boats. By May 1943 aircraft, vessels, and procedures were well in place, and HF/DF was being used offensively to good effect. Capt. Peter Gretton, one of the most successful convoy-escort commanders in the Royal Navy, called shipboard HF/DF sets "invaluable devices" that not only gave warning of impending attacks, but could indicate from which direction the attack was most likely to come.[61] At last the means were available to take the initiative away from the U-boats and to attack them before they could strike. In fact, "High frequency direction finders were the only long-range shipborne submarine detectors available to the Allies during the war."[62] They could accurately and quickly locate U-boats when they were still too far away to launch torpedoes but close enough to be attacked by air- and seaborne escorts. Pressed by the heavy burdens of a drawn-out war, the Royal Navy haltingly developed the somewhat primitive FH3 HF/DF device and then used it with astonishing skill and to very great effect.

In the meantime a French scientist from ITT had offered the U.S. Navy a more sophisticated apparatus—the DAQ, equivalent to the British FH4. Finally convinced of the significance of shipborne HF/DF, the U.S. Navy began to churn out DAQs at a great rate. The navy then followed British experience in using HF/DF against the U-boats.

CHAPTER 5

The Inventor: Henri Busignies

Toward the end of 1928 a recently created ITT Paris subsidiary, the Laboratoire Téléphonique, was looking for additional staff. A young engineer named Henri Busignies walked into the office of Maurice Deloraine, the lab's director. Busignies offered Deloraine and ITT a license under a patent he had recently obtained, covering a direct-indicating radio direction finder.[1]

At first Deloraine was highly skeptical because he knew from experience that most inventors tended to overestimate the significance of their inventions. He assumed that it would be difficult to come to any kind of agreement with the young engineer. After examining the patent cautiously, Deloraine explained that while it did seem to have some merit, it needed a great deal of work before it would be ready for commercial production.

To Deloraine's surprise Busignies agreed with him, saying he was willing to sell his patent rights for a nominal sum. What he was most interested in, said Busignies, was having the opportunity to continue the development of his direction finder with adequate financial backing and support. Impressed, Deloraine quickly agreed on a purchase price, and Busignies became a member of the Paris laboratory. He continued to work for ITT until his retirement as chief scientist emeritus in 1975.[2]

Later in his career, Henri Busignies escaped from France under the nose of the Nazis and took his direction finder to the United States, where it helped to swing the Battle of the Atlantic in favor of the Allies. After the war, his innovations in aerial navigation played an important part in assuring safety for commercial flights. Ultimately, Busignies could claim credit for more than 140 important inventions.[3]

Dr. Henri Busignies, approximately 1939.
Courtesy of Cécile Busignies

Henri Gaston Busignies was born at Sceaux, near Paris, on 29 December 1905, the son of Henri Busignies, a mechanical engineer from the north of France. By the time he was fourteen Busignies was a radio ham, and, as he himself expressed it, "right from the first I was playing with crystals and tubes."[4] But Busignies soon realized that he was more interested in experimenting with new radio circuits than in picking up transmissions from New York or Sydney.

In 1926, after studying at the Jules Ferry College in Versailles, he obtained a degree in electrical engineering from the Institut Normal Electro Technique in Paris. While keeping up with his assigned work, Busignies continued to develop an early flair for circuit design; in addition to his regular studies, he found time to conduct experiments of his own. While still a student he obtained a patent on a radio compass, a device that electronically "points out" the direction of radio stations. Eventually this patent was to be followed by almost a hundred others.[5]

After completing his military service, Busignies went to work for a laboratory in Paris where he quickly became the expert in avionics—the science of electronics applied to aeronautics. When Charles Lindbergh landed in Paris after his epic solo transatlantic flight in 1927, he was having trouble with his compass, which had been malfunctioning. He was directed to take it to a young engineer, Henri Busignies, who was able to correct the problem. Busignies's career in radiolocation was well launched, as was his connection to the United States.[6]

The following year Busignies went to see Maurice Deloraine at ITT.

March 9, 1943. H. G. BUSIGNIES ET AL 2,313,047

DIRECTION FINDING SYSTEM

Filed Aug. 1, 1941

FIG. 1.

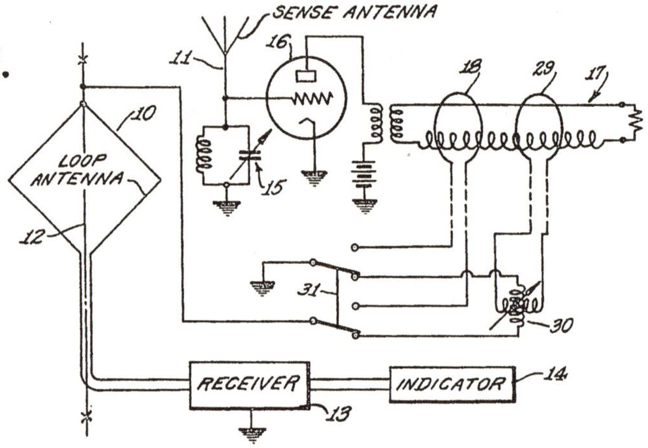

FIG. 2.

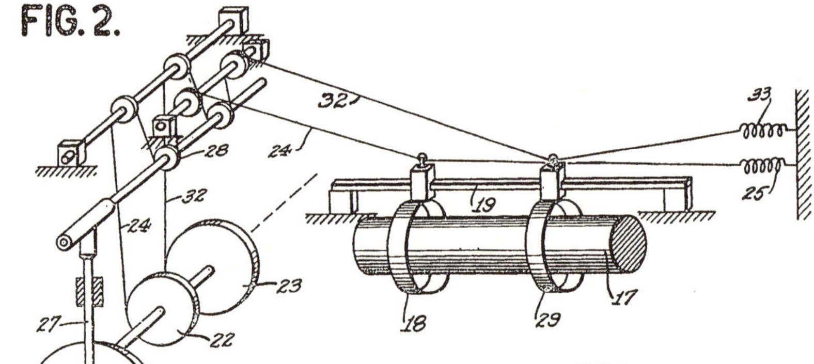
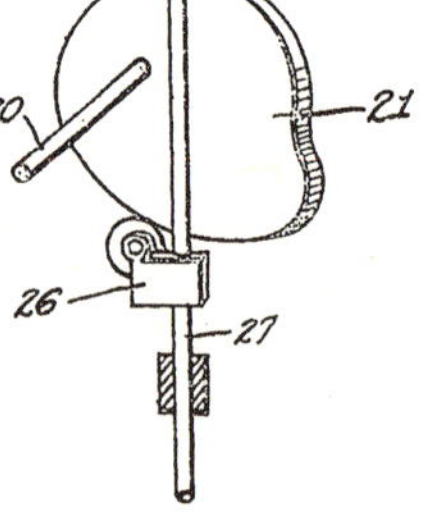

FIG. 3.

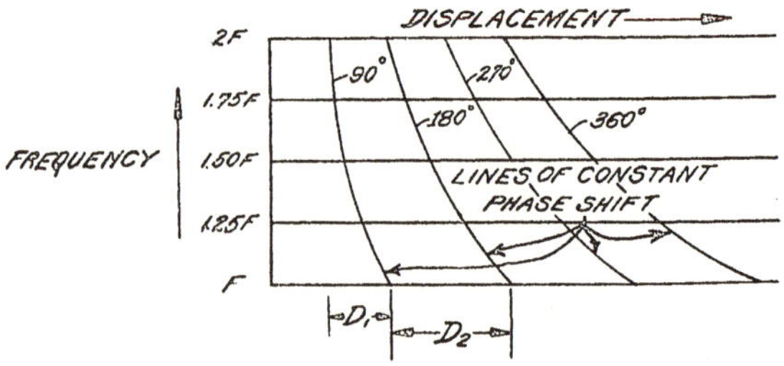

INVENTORS
HENRI G. BUSIGNIES
AVERY G. RICHARDSON

BY

E. D. Phinney

ATTORNEY

A small division of experts on direction finding was formed under Busignies's leadership, and soon they became the most experienced in France in this field. For the next twelve years, in addition to radio direction finders, Busignies helped to develop airplane navigation systems and early radars.[7]

Henri Busignies was five feet, ten inches tall, and slightly built. He was quiet, soft-spoken, and serious, remembers Ruth Lockhart Lombardi, his secretary and later technical assistant for twenty-one years at ITT. Busignies was totally dedicated to his work, which he enjoyed enormously, and he went about it with such intense single-minded concentration that "most of the time you didn't even know he was there."[8] In fact, he seldom took time off from work, even on weekends. "Ruth [Lombardi] used to say she hated to see him Monday morning," recalls Mme Busignies, because "he was writing so much on the weekends."[9] In the 1950s he was rewarded for his many contributions to electronics with two honorary doctorates, one in science from the Newark College of Engineering and one in engineering from the Polytechnic Institute of Brooklyn.

As might be expected of an inventor, Busignies was a rather solitary person, though he was also a devoted husband and father. He was generous to those who worked for him, and one assistant, Avery G. Richardson, has written that Busignies "hated to say no to anyone's ideas; he would just sit back and inspect the ceiling light fixtures." In this way he "incited the creative minds around him," so that they would return to him next time with their ideas better formed.[10] Sloppy apparel and bad language irritated Busignies, but the only thing he really did not have time for was people who were not very bright. Perhaps this was inevitable, since he has been described as so brilliant that he had "a telephoto lens on the future."[11]

In many ways Busignies's career and interests paralleled those of Robert Watson-Watt. As superintendent of the radio department of Britain's National Physical Laboratory, Watson-Watt had been involved in intensive investigation of direction finding for many years prior to World War II. In 1926 he published a speculative paper on "An Instantaneous Direct-Reading Radiogoniometer." But it was Busignies, later called "one of the world's most prominent scientists in this field," who first succeeded in making the device a

Fig. 5.1. H. G. Busignies, et al., 1943 patent for direction-finding system.

Mme Cécile Busignies, in a photograph taken in France.
Courtesy of Cécile Busignies

practical reality while others only theorized.[12]

Between world wars there was a rather free international exchange of scientific information, both in scholarly and professional journals and at conferences. Early British work on direction finders had yielded inaccurate results at high frequencies from shore-based equipment, and the results from mobile shipboard systems seemed practically hopeless. So when Busignies's automatic radio direction finder (or radio compass) successfully guided a plane all the way from Paris to the island of Réunion off Madagascar in 1936, the dramatic public display attracted a great deal of attention.

The next year Busignies was demonstrating his equipment in the United States courtesy of the Paris Laboratoire's parent company, ITT.[13] While in New York the Frenchman was invited to give a lecture on his direction finder to the Radio Club of America. As he told that same body thirty-nine years later, after the lecture he was treated to "a first-class dinner—for eighty-five cents!"[14]

The performance of the French radio direction finder was widely publicized, and Busignies continued to work on developing it through the early months of the Second World War. But it seems that the German U-boat Command must have been more influenced by the early weaknesses of the equipment than by its steady improvement.

In the meantime Busignies's first outstanding success after the Réunion flight was the introduction of his direct-reading radio compass in the French air force. As this apparatus is also referred to as a direction finder, there is some confusion in terms. The main difference between a radio compass and a direction finder is one of use—are you inter-

Ruth B. Lockhart (Lombardi) with Henri Busignies at the ITT offices in New York, early 1970s.
Courtesy of Ruth Lockhart Lombardi

cepting a signal in order to figure out where you are or in order to find the source of the transmitter? With a radio compass the point is that you already know the source of the transmitter. It is transmitting a signal from a known location so that you can establish your own position in relation to it. A radio compass, therefore, works on cooperative transmissions, unlike a direction finder, which has to be able to locate the source of uncooperative transmissions.[15]

Busignies's first automatic direct-reading radio compass (which was eventually presented to the Smithsonian Institution in December 1975) was developed in the 1930s for airborne use, where radio bearings must be taken very quickly because of the speed at which planes travel. It was put into service by the French air force for navigational purposes.[16] At this time the French navy too sought technical assistance from ITT. The navy was considerably behind the British in the development of asdic, which in 1938 was still in the laboratory stage. The hydrophones were effective, however, and they were especially up to date in radio technology.[17]

Busignies's HF/DF story really started in 1938, when the French navy

asked the Paris laboratory, because of its established reputation in the field of direction finding, to develop what was called an "instantaneous" radio direction finder. The automatic direction finders so far developed indicated the direction of arrival of a radio wave on a 360-degree dial. The operator only needed to tune to the transmission frequency to read the bearing of the distant radio transmitter from the indicator. But the indicator scale took a few seconds to settle into position. If the radio signals were very short, the scale could not give a correct bearing.

The problem raised by the French navy was to find a method whereby the bearing would be shown correctly even if the transmitted signal was very short, lasting a second or less. It seems clear that the navy had somehow obtained information indicating that German submarines were planning to use extremely brief coded messages transmitted in the high-frequency range. The theory behind this German method of transmission was that no direction finder could get a bearing on such short signals; the position of the transmitting submarine in consequence could not be established.[18]

Busignies and his colleagues at the Paris laboratory were guided by information from the navy in their work on a direction finder capable of pinning down the location of the originating transmitter of these very short HF signals. Eventually suitable designs were produced, and by May 1940, when the German armies invaded France, four models of instantaneous, automatic bearing indicator, visual, HF direction finders had been completed and tested.[19]

This design involved several new concepts. One longstanding weakness in direction finding had been the difficulty of dealing with polarization; that is, the 180-degree origin problem that indicated a line of bearing, but not from which end of the line the transmission had originated. It was this problem that prevented the development of shipboard HF/DF in the Royal Navy until 1941, when W. Struszynski created a reliable sense antenna.

Busignies designed a brilliantly innovative antenna construction that reduced polarization errors to a minimum, with an HF goniometer revolving at something like twenty revolutions per second, and a magnetic coil rotating synchronously with it around the

Henri Busignies with parts of his airborne automatic (low-frequency) radio compass.
Courtesy of Cécile Busignies

neck of a cathode-ray oscillograph. The direction from which a signal was received was shown by the position of a propeller-shaped pattern on the cathode-ray tube. This system was so fast in response as to be instantaneous for all practical purposes.[20]

The addition of the cathode-ray tube as an indicator, instead of the moving scale, made it possible for the first time to establish a bearing on the briefest possible transmission. The instantaneous type of indication was given on a small round screen treated with fluorescent material. When an impulse caused by a radio signal was received it excited the screen, leaving a trace (like the spot of light that slowly fades when a television is turned off) from a signal lasting as little as one-tenth of a second.

Avery Richardson, one of the American engineers who went to work for Busignies in New York City in early 1941 and who was already familiar with his work, has described Busignies's new contributions thus:

> While I had been making airborne D/F with a left-right indicator, he had specialized in cathode-ray indicators for a very good reason. The simple, rugged meter indicator [which Richardson had been using] is fine for peacetime navigation using beacons and broadcast stations. But Busignies knew the Germans were using very short signals the meter will not handle. So he rigged up a motor-driven goniometer giving thirty bearings per second and displayed them on a cathode-ray tube. This was a very cute trick to get high-speed bearings. Bearings had previously been taken by manually rotating a large loop to get an audible null signal; but an 1800-rpm loop? The Bellini-Tosi goniometer was old but large. What Busignies did was to miniaturize it.[21]

The use of a fluorescent screen also had the great advantage of showing more clearly the quality and types of signals received. As much short-wave propagation is due to ionospheric reflection, the display furnished considerable information on this quality in a manner that had never been available before. This was to prove particularly significant when HF/DF was considered for tactical use aboard ships. A signal received from a sky wave might have originated a thousand miles away, while the source of a ground-wave transmission would be less than thirty miles distant. Busignies's device made it easier to tell which kind of wave was being received.

As director of the Laboratoire Téléphonique, it was Dr. Deloraine's responsibility to maintain close contact with the French military and civil authorities who were the market for Busignies's direction finder. Busignies's work was also very much of international interest, especially in the field of navigation. Since his laboratory was a subsidiary of ITT, Deloraine had extensive connections in the United States as well as in England and in the many other countries where there were foreign affiliates of the parent company.

E. Maurice Deloraine was born in Paris on 16 May 1898. In 1916 he began his university studies at the Ecole Supérieure de Physique et Chimie, the highly competitive institute made famous by the work of Pierre and Marie Curie. In 1917, their studies interrupted by the war, Deloraine and his fellow students were sent in a group to the French army signal corps. Deloraine has written engagingly about his adventures with the haphazard radio-telephone communications on the western front. In 1918, after the war ended, he obtained the equivalent of a B.S. degree, the Certificat de Mathématiques. In 1920 he received a diploma in engineering from the same Ecole Supérieure, which was a branch of the Université de Paris. In 1949 Deloraine was granted the degree of doctor-engineer by that university.[22]

Moving to England in 1921, he became a member of the engineering staff of the International Western Electric Company, soon to be acquired by ITT. In London, Deloraine worked on radio broadcast transmitters and was responsible for part of the development in Great Britain of the first transatlantic radio telephone circuit.[23]

During the next years he worked closely with Sosthenes Behn, the founder and president of ITT. Behn was always referred to as Colonel, having attained that rank in the U.S. Army Signal Corps in World War I. Though he was an American citizen, Behn was also a true internationalist and an early proponent of the multinational corporation. Under his leadership ITT spread into Great Britain, France, Belgium, Spain, Italy, Holland, Austria, Hungary, and finally, after 1929, Germany. Deloraine, Busignies, and others on Behn's staff made frequent trips to each of these countries.[24]

Such internationalism was not well understood in the United States, and when war broke out it may have led to suspicions with regard to the allegiance of the Frenchmen. This was to have an unexpected effect later, when Deloraine and Busignies were in the States. Robert Murphy, U.S. ambassador to France from 1930 to 1945, wrote a preface to Deloraine's 1976 book of reminiscences about ITT in which he pointed out:

Sosthenes Behn and his associates [played] a vital role in the highly important telecommunications field. These men comprised several nationalities and they seemed to work effectively as a team. Of course there was no question in my mind regarding Mr. Behn's patriotism and loyalty as an American citizen. But I did not foresee that during World War II the technical knowledge of ITT staff members would make extremely important contributions to our national defense, as demonstrated by their development of the instantaneous direction finder.[25]

Apparently Ambassador Murphy felt it necessary to state that he had never doubted Behn's patriotism. After all, he must have known that the colonel had risen to be assistant chief signal officer in the First Army in the Great War. Behn had even been awarded the Distinguished Service Medal and membership in the French Legion of Honor. Yet the patriotism and loyalty of ITT's French engineers was indeed questioned, and initially this may have reduced their ability to contribute to the war effort. In fact, the work of Behn's Frenchmen during the war was constrained by both British and American suspicions.

On 3 September 1939, with the implementation of mobilization plans in France, the Paris laboratories of ITT came under full military control and became a part of the national defense services. The personnel were all mobilized, all research and development expenses were paid by the government, and the entire work program was dictated by military considerations.[26]

As part of this war reorganization, Deloraine assumed that there would be cooperation with the British military establishment in the matter of scientific exchanges, especially with regard to radar, radio communications, and direction finding. Because of his early years working in England and the literally hundreds of visits he had made there since, Deloraine was particularly aware of how such cooperation could help in the French war effort. He stated categorically in his reminiscences that the British navy had been kept informed of the development of Busignies's HF/DF as well as of other research in progress at the laboratory. And Colonel Behn himself had encouraged full cooperation between his British and French laboratories from the moment of the outbreak of war with Germany.[27]

In fact, there had been technical exchanges between the British and French navies for some time. These contacts had begun in response to the Italian invasion of Ethiopia in 1935 and the subsequent British interest in access to French naval bases in the Mediterranean. Technical conversations between all service branches of both countries took up again in March 1936 when German troops marched into the Rhineland, and exchanges stepped up in 1939 as part of the meetings of the Anglo-French general staffs in March, April, and May.[28]

Noting that early in 1939 the British Committee of Imperial Defence had decided to share even "most secret" equipment with the French, Robert Watson-Watt participated in various of the staff conversations. By then he was the scientific adviser on telecommunications to the Air Ministry and the recognized head of British radar. Watson-Watt met the French air attaché in London in April, and he met members of French commander in chief General Gamelin's mission early in May. Later that month he was in Paris comparing notes with the French air force on ultra-high frequencies and direction finding. Watson-Watt recorded that he found "little to admire" in France, and that the development of a high-powered pulse system of detection had only just begun in the labs of the Sadir company (Société Anonyme des Industries Radioélectriques).[29]

In August there was an exchange of French and British liaison officers at several major naval bases, and on 25 August a highly secret Anglo-French naval signal code was put into effect. By 3 September 1939, when war was declared, British naval officers were already working in the French navy ministry in Paris.[30] But according to the vivid account of Maurice Deloraine, when Robert Watson-Watt visited Paris that fall, he informed an astonished Deloraine that "the question of security in France had not been solved satisfactorily, and that in consequence they could not give us access to their techniques." Deloraine further maintained that this was later confirmed via official channels.[31] Indeed, it was not unlike the similar early British reluctance about sharing scientific know-how with the United States.

For his part, Watson-Watt also recorded a visit to Paris at the beginning of November, to select sites for radar stations that were to form an early warning chain (like the British Chain Home network) from the Pas de Calais to the Maginot line. Watson-Watt does not mention speaking to Deloraine on this visit, though he did have a day of conferences in Paris that left him with the "leaden impression that the French war-machine was more sensitive to obstacles than to opportunities."[32]

A similar story to that told by Deloraine has been noted by Capt.

Paul Auphan, who was serving on the staff of Adm. Jean Darlan, head of the French navy. That same November, 1939, the British first lord of the Admiralty, Winston Churchill, accompanied by Admiral Sir Dudley Pound, visited Darlan's wartime headquarters at Maintenon, sixty-five kilometers west of Paris. After a cordial meeting in which both sides expressed total agreement on the need to work together with the sole aim of winning the war, Churchill, as he was leaving, told Darlan that he had complete confidence in him and his staff. But, the Englishman continued politely, that confidence did not, unfortunately, extend to the navy ministry or to the French politicians. Would Darlan please not keep those gentlemen too well informed of current secret operations, as they seemed incapable of "holding their tongues."[33]

Referring to his impressions from his own November visit, Watson-Watt also mentioned visiting Admiral Darlan. It was obvious to the Scotsman that Admiral Darlan "certainly had mental defences against *Perfide Albion* as well as against *Sale Boche*."[34] All of these stories indicate a strong sense of British distrust of French determination for the fight at hand, and also of French awareness of this attitude. Deloraine was left with the feeling that as far as technology was concerned, Britain and France were fighting separate wars. Watson-Watt, on the other hand, believed Britain was sharing all it had while the French were unable or unwilling to make effective use of it.

In Deloraine's account of these anxious months of the "Phony War," British obtuseness finally ended in March 1940, when a group of scientists from London visited the Paris labs and informed him that the previous decision not to allow cooperation had been reversed. "We come late, but repentant," they reportedly told Deloraine.[35]

Of course, they did not know then how late they really were. But since the Frenchmen had now apparently been cleared, they were invited to visit Britain and examine installations there. Deloraine records that they visited the Royal Navy research establishment at Portsmouth and several army research centers, as well as various radar installations along the coast.

The French engineers returned to Paris with valuable technical information. In exchange they had disclosed their own "various developments" to the British. Unfortunately Deloraine does not specify whether or not this included Busignies's new instantaneous HF direction finder. Since that was the most critical device that they were then completing, and since Britain would clearly be facing the brunt of the maritime war against Germany, it seems unlikely it would have been withheld.[36]

Watson-Watt also noted Deloraine's visit to England that spring. He observed:

> I cannot now remember when it was that I had a brief encounter with General Gamelin, nor when it was (in April? 1940) that I showed Maurice Deloraine (of Le Materiel Telephonique [*sic*]) our chain stations at work. But I do know that it was Gamelin's elaborately tended and preternaturally blond coiffure, and Deloraine's air of cool detachment that gave me early warning that France would fall. "This time," I told myself, "*Ils ne tiendront pas.*"[37]

How much Watson-Watt's Francophobia affected the exchange of information about Busignies's HF/DF it is impossible to say, but he was very influential when it came to British telecommunications developments. It is also possible that Watson-Watt's latent distrust was so obvious that Deloraine withheld his HF/DF information. At the very least, such memories may have colored Deloraine's account of his exchanges with the British scientist in the fall of 1939 and the spring of 1940.

Writing in 1976, nineteen years after Watson-Watt published his own wartime recollections, Deloraine may not have forgotten that the Scotsman only mentioned him one more time in his book. Often referring to himself in the third person, and sometimes only by his title, Watson-Watt had recorded that if he had not been in New York on 12 and 13 February 1942, talking to, among others, "that same Maurice Deloraine who had silently dispirited him [Watson-Watt] in 1939," he might have been on hand to solve one of the minor radar glitches responsible for the humiliating dash up the Channel of the German battlecruisers *Scharnhorst* and *Gneisenau*.[38]

Somehow Deloraine's name is even linked by Watson-Watt with a demoralizing British defeat at the hands of the German navy! Indeed, there seems to have been a continuing sense of hostility between the British and French scientists, which, in addition to infecting their memoirs, may well have inhibited their professional cooperation in the stressful early months of World War II.

Those were the very months when the British temporarily shelved shipborne HF/DF in favor of radar, and it may well be that they dismissed talk of the French device, assuming it was no more effective than their own. It is a provocative thought that shipboard HF/DF capable of producing accurate bearings might have been introduced as early as fall 1940, had the British and French been more adept at sharing their newest technologies.

The case of radar has some interesting parallels to that of HF/DF. Having shown Deloraine the status of radar development in Britain, the British authorities stressed the secrecy surrounding the technology. This put Deloraine in an awkward situation, since he knew that most of the fundamental radar concepts had already been openly published. In consequence, as soon as he returned to France he asked his director of patents, "who was very good on documentation," to search out all published material relating to radar, in any language.

For example, radar had been operating in the French ocean liner *Normandie* before the war, and a detailed description of it was easily available in print. The *Normandie* radar even had a pulse magnetron, which the British considered a technology uniquely theirs. The Tizard Mission was to present it to the United States with the greatest secrecy in September 1940![39] Many radar patents of various types had also been published, including those covering the principles of PPI (plan-position indicator), developed in the Paris lab, and MTI (moving-target indicator), which was invented by Busignies and was later in worldwide use at airports for navigation and traffic control. Foreign publications with articles on radar, including some from the United States, were numerous.

Deloraine assembled a sizable bundle of these journals and sent it to England, pointing out that secrecy could only be invoked as far as additional techniques were concerned. It seems that the British were surprised when the package arrived, but they requested that the information in the journal articles just sent to them should be classified as secret anyway! Deloraine, though amused, complied. He realized that, in any case, the war was just about over for France.[40]

With the breakup of the Sedan hinge in the fortified Maginot line, it seemed clear to many in France that the country would fall to the Nazis. Deloraine still had to operate in accordance with military orders, but this was not particularly inhibiting as he usually wrote those orders himself and then got them stamped and signed by the proper authorities. In this way ITT Paris began to move files, drawings, models, laboratory notebooks, and some of its personnel south to a location that had been earmarked for this purpose in case of necessity. All papers left in Paris were burned. Finally the French general in charge told Deloraine to evacuate everyone still in the Paris laboratory, and on 12 June 1940 the last ITT people left.[41]

When the German armies arrived in Paris forty-eight hours later, the ITT laboratory was empty of personnel and equipment. Some of its models were still in a number of railway cars scattered over the country. The only four existing models of Busignies's instantaneous direction finder had been disassembled and dispersed, and they were never reassembled during the occupation.[42]

Some of the confusion of those weeks, and the understandable suspicion that it aroused in Britain and later in the United States, emerges in the next stage of Deloraine's remarkable saga. Though the British may have been slow to acknowledge the value of the French scientists' cooperation in the war effort, it appears that once they made up their minds, they were ready to do whatever might be necessary to protect their French associates. Immediately after fleeing Paris Deloraine was summoned to Bordeaux, the temporary new seat of the French government. Arriving, he was informed that French headquarters had received a telegram from the British Admiralty stating that the personnel of ITT laboratories could embark on any available British ship. Deloraine was also informed by his own government, however, that he and his associates were still mobilized. If they took advantage of the British offer they would have to face the consequences of desertion from France in time of war.[43]

Since their families were scattered and they would not leave without them, and as the idea of desertion was not appealing, the French engineers remained. On 17 June, the French authorities asked Germany for an armistice, and a demarcation line was established in France between an occupied and a so-called free zone. Deloraine and his colleagues decided to split the laboratory in two, establishing at least half of its personnel in the non-occupied zone. They rented space in Lyons and reassembled there all the equipment that had not been lost. Meanwhile the occupying forces had noticed the absence of activity in ITT's Paris laboratory, and the French scientists were invited to return without delay. At the insistence of the German authorities, the ITT labs and factories in France were reactivated at the end of July 1940.[44]

Deloraine, Busignies, and some of the others returned to their posts in Paris, where they hoped they might still be of some use to the anti-Nazi effort. They tried to assume the unobtrusive guise of scientists engaged in general research. But because of their international reputations and the extensive business they had been doing abroad, particularly in Germany, it was not easy to maintain the charade. Deloraine was accosted by a German army colonel, formerly an engineer with the Deutsche Post, whom he had met at an exhibition before the war. The colonel had been sent to find out what had happened to all the equipment from the lab. When Deloraine showed him the military evacuation orders, signed before the armistice had taken place, the colonel was satisfied: orders were orders. After that no serious effort was made to find out what the French scientists had been up to in their lab.[45]

Busignies had a similar experience when he was confronted by a Luftwaffe colonel who had previously known him and who knew of his expertise in military electronics. The colonel expressed interest in learning of Busignies's research activities and suggested that they get together for dinner. Only by some really fancy footwork was Busignies able to evade that and subsequent invitations.

The threat of German interference was present all the time. There were guards at the laboratory, and guards followed the scientists and even their wives to their homes. "Even at the door of our house, I had Germans," recalls Mme Busignies.[46] Probably this was due less to interest in the work that was going on at ITT than to the wish to ensure that the Frenchmen would not leave the occupied zone. Deloraine did not think the Germans were after his scientific know-how. "It appeared to us," he wrote, "and this was confirmed later, that they felt they already had all the necessary technical knowledge and specialized equipment to defeat England."[47]

The Germans were close to laying their hands on the HF/DF device that was to play an important part in their eventual defeat. But it seems that their assurance of scientific superiority was such that they even neglected to see if there was anything to be learned from an engineering laboratory they knew had been turning out advanced military equipment in peacetime. Meanwhile the ITT executives in New York were extremely concerned about the fate of their French scientists and the work they had accomplished. In September 1940 a message from Colonel Behn was passed to Deloraine via the United States embassy in Paris, suggesting that as a first step Deloraine and a couple of colleagues should try to escape to North America. At this point it was not the U.S. Navy, as has been suggested, but ITT executive H. C. Bohle who arrived in Vichy to help facilitate the departure from France.

But by then it was mid-October, and it was no longer easy to pass from occupied to unoccupied France. An Ausweis, a permit from the German authorities, was required. Fortunately a French police official known to Deloraine obliged by obtaining the permits—for a fee already well established with the occupying forces.[48]

Eventually a total of eleven persons made their way separately to a secret rendezvous in Lyons in unoccupied France. In addition to himself, Deloraine took Busignies "because of his exceptional knowledge in direction finding; Labin, as our principal expert in radar; and Chevigny, a vacuum tube expert," as well as their wives and children.[49]

The engineers were not content to get themselves out of France empty-handed, however. Before leaving Paris they reproduced as many laboratory drawings as they could. Then they wrapped the drawings, and vital parts of several working models, in six or seven bundles of brown paper and handed them to a train guard in the Paris railroad station. The guard, who was not in on the secret, agreed to place them in a corner of his railway car with the other baggage, providing someone in Lyons would be there to pick them up when the train arrived. The bundles eventually reached the U.S. embassy in Lyons, which refused to send them directly to the United States but agreed to send them on to Lisbon, in neutral Portugal, where the engineers finally caught up with them.[50]

The usually meticulous Nazis were in the habit of searching all the trains leaving the occupied zone, but, fortuitously, in this case they examined everybody and everything except the baggage car at the end of the train! At the risk of their lives, the French ITT scientists smuggled out of occupied France the plans for a technical device that was to rank with radar and sonar in the defeat of the U-boats.

Although her husband did not know it until later, Mme Cécile Busignies had already been risking her life regularly since the beginning of the German occupation. She had been acting as a courier for the Resistance, traveling all over France delivering clandestine messages. Mme Busignies's code name was Moustique Effrayant, or "Scary Mosquito," because, as she explains, her enthusiasm and energy for the cause were so great that "even before they told me what to do I was already running . . ."[51]

From her home in Saint-Cloud, a suburb of Paris, Mme Busignies would set out sometimes accompanied by her young daughter, Monique, on the pretext of visiting relatives, perhaps her parents in Chartres. In her very modest way, and un-Americanized after fifty years in the United States, the diminutive, vivacious Mme Busignies describes hair-raising assignments. These involved rides hidden under hay in wagons, a meeting with a priest "who was no more a priest than I or you," and trips to escort Jews to the safekeeping of some valiant nuns in a convent. The Jews "were from everywhere," recalls Mme Busignies, "some from Russia, some from Poland, and some from Germany . . . They had been living in Paris for some time. And [now] they were afraid to be deported."[52] Even when she and her family had escaped to Lyons in Vichy France, Mme Busignies's work was not over. She continued to carry messages to Marseilles and other places along the south coast, including, on occasion, Italy.[53]

Meanwhile the whole party from the Paris lab was delayed in Lyons for some time, while Mr. Bohle from ITT tried to arrange for their departure. They received little encouragement from the U.S. embassy. American officials believed that the Vichy government was set to collaborate with the German authorities and they should not, therefore, facilitate the exit of a group of French engineers specializing in matters relating to military equipment, whose loyalty to the Allies might be suspect.

Just as in Paris, however, it turned out to be a matter of many people being willing to put one over on the Germans—for a small fee. As Mme Busignies put it, "Not everybody was bad in Vichy. We got the right papers. I think that under the Germans they developed lying to the Germans."[54] Still, Deloraine, ever prudent, did not want to leave France without some assurance that he and his party would be able to enter the United States.

He need not have worried. Colonel Behn intervened with the State Department, and Deloraine himself saw a telegram from "Sumner Welles, Acting," instructing the consul at Lyons to issue the appropriate papers. Deloraine remembers the telegram was "in clear," which made him nervous. The Americans did not seem to realize that this was really war.[55]

The wanderers next moved, in a succession of hops reminiscent of Ingrid Bergman's experiences in *Casablanca,* to Marseilles, Algiers, Casablanca, Tangier and Lisbon. They traveled by seaplane and by train, all the while aware of the danger of exposure. General Francisco Franco,

Spain's Fascist leader, "didn't want anybody crossing [from French to Spanish Morocco], and he said, 'if they cross, I put them in jail.'"[56] So the Frenchmen and their families split up into small groups again, to avoid attention, and made their separate ways to Portugal. The Busignies family was lucky and rented a small private plane. Mme Busignies accomplished her last mission for the French Resistance on the trip to Casablanca by wearing a fur coat that had a secret message sewn into the lining. This might easily have been her last mission altogether, as Mme Deloraine, who was accompanying them, kept marveling at the coat, insisting loudly that she had not known Cécile had a fur, particularly one that was obviously much too big for her![57]

With the group all together at last in Lisbon, except for the Chevignys, who joined them later, they left on the old American tramp steamer *Siboney*, which seemed about to fall apart. They were traveling in good company, however, as in addition to a number of European Jews fleeing the Nazis, the passenger list included the famous French aviator and writer Antoine de Saint-Exupéry and the film director Jean Renoir. Also aboard was an old American friend, Fred Caldwell, head of CTNE, an ITT affiliate in Spain.

Fortunately the precaution was taken of placing the precious laboratory drawings in Caldwell's cabin. The *Siboney*, after a rough passage, put into Bermuda, where the British inspected the cabins of the French engineers very carefully. Here the Frenchmen received the distressing news that they had been classified as "enemy aliens" by their British "allies," and had the drawings been found they would certainly have been confiscated. Nobody searched Caldwell's cabin.[58]

After a six-week journey the small French party finally landed at Hoboken, New Jersey, on 31 December 1940. Looking very much the part of bedraggled refugees, the French engineers and their families took the Hudson subway line under the river "to arrive late in Manhattan," remembered Deloraine, "in the middle of what appeared to us the most extravagant and uncalled-for year-end rejoicing."[59]

The very next week Colonel Behn, always in a hurry, arranged a meeting between his French engineers and military personnel in Washington. The rank of the Americans present shows the extent of Behn's influence. Rear Adm. Leigh Noyes, the director of naval communications, was there, as well as Brigadier General Mauburn, his opposite from the army. Also present was Lt. Comdr. Frederick R. Furth, Noyes's technical adviser and the officer in charge of fleet communications and equipment (including radio, radar, sonar, and electronic countermeasures), and Col. Roger Colton, head of the Fort Monmouth laboratory, who was taking notes. In all there were about twelve people seated at a long table, and Dr. Deloraine was invited to address the group.[60]

Deloraine confined his remarks to a description of the technical drawings he and his colleagues had brought with them from France, along with an explanation of the performance that could be expected from the equipment, if produced. A number of questions were asked, and in the afternoon there were further meetings that probed more deeply into the capabilities of the ITT devices, particularly the instantaneous direction finder.

The Frenchmen were a bit disappointed by the meetings, as they felt the Americans were cagey—while pumping them for information, they seemed very careful not to give anything away themselves. But a week later that all seemed to change, when Comdr. LeRoy Blaylock from the Bureau of Ships arrived at the New York offices of ITT, at 67 Broad Street in Manhattan. Commander Blaylock told Deloraine that he had "caught hell" from his superiors for not being up to date with technical developments. "If you want a contract," he continued, "you have it, but you'd better be ready to deliver what you promised."[61] Industry was well ahead of the navy in D/F technology, and only Colonel Behn's dynamism had forced the navy to sit up and take notice.

Commander Blaylock's visit was quickly followed by the arrival of a "Letter of Intent" to produce, within the shortest possible time, an experimental model consisting of four instantaneous direction finders operating over the shortwave band. This came to be known as the model DAJ high-frequency direction finder for shore stations. Since Busignies had not only assisted in smuggling the technical drawings of his direction finder out of France but had also carried out vital parts of the actual device hidden on his person, the Frenchmen were able to get off to a fast start in its production.[62]

The U.S. Navy's Huff Duff program really took off with the development of the ITT DAJ. In 1976 Dr. Busignies donated the original prototype to the Smithsonian Museum, and a curator writes that the museum "now has the experimental HF radio direction finder . . . which Busignies developed ca. 1936 and brought to this country during WWII. From this apparatus a system was apparently developed by the navy for both coastal station and destroyer escort HF/DF."[63]

It took only three months for ITT to fulfill the conditions of the first navy contract. In one day, Deloraine selected and rented fifty acres at Great River near Amagansett, Long Island. He and his team installed their equipment in wood shacks there and set to work. The great speed with which the four Busignies high-frequency direction finders were produced attracted favorable attention.

In April Commander Furth, accompanied by several other officers, visited the new facility. A few days later a larger delegation appeared, including various admirals and a group from the National Defense Research Committee (NDRC). The secretary of this committee informed Deloraine that several months earlier the committee had been charged with solving the problem of developing an HF/DF capable of obtaining usable bearings from very short high-frequency transmissions, but that they had not progressed very far as they had, in effect, started the project from scratch.[64]

If Deloraine's account is accurate, the NDRC apparently had not consulted the Radio Direction Finder section of the Naval Research Laboratory where Dr. Maxwell Goldstein and his team had been working on HF/DF for several years. This is curious as Goldstein himself was the naval representative on the NDRC, though that was during the war, so it is possible that he had not yet been appointed to the position. Deloraine's account would also seem to indicate that there had been no significant exchange of D/F information between the United States and Britain up to this time.[65]

What happened next, however, is something of a mystery. It seems there was a six-month hiatus, or at least there is not much on record of further navy interest in Busignies's D/F until October 1941. The four models already ordered were being tested, and perhaps it was felt expedient to await confirmed results.[66]

There was no enthusiasm to develop shipborne HF/DF, which some still feared would necessitate serious curtailment of the firing efficiency of a vessel, because of the special antenna masts. An NRL report of June 1942 admitted: "This incompatibility of naval aims has probably been responsible for the discouragement of shipboard high frequency direction finding in the past."[67] In a more general way, too, it has also been widely suggested that Americans had not taken the U-boat threat seriously until the spring of 1942, when the Germans began to sink ships in great numbers directly off the East coast.

What is certain is that there were doubts about the allegiance of the

Frenchmen. These doubts emerged at this time, and they may have been responsible for putting a brake on the development of HF/DF for several months. After the war it was discovered that the arrival of the engineers as a group, complete with drawings and bits of actual equipment, was regarded with suspicion by military security and by the FBI. It was felt that the Frenchmen might be German "plants." This possibility was taken so seriously that notice was sent to all military departments to "proceed with extreme caution" in any contacts with the French ITT engineers. This notice was even circulated to civilian concerns such as Bell Laboratories, where many former friends of Deloraine and his group kept well away from them. "Some others," notes Deloraine, "completely disregarded the advice."[68]

In fact, narrow chauvinism was not just the understandable product of the war emergency. It was an attitude, thoroughly intertwined with a keen sense of business competitiveness, that grew steadily in the United States in the 1920s and 1930s, particularly with regard to the young radio electronic and telecommunications fields. By the close of World War I Franklin D. Roosevelt, assistant secretary of the navy, had become convinced that foreign ownership of any part of the American telecommunications industry would be extremely dangerous in a future war. As president, Roosevelt was able to act on this conviction when he facilitated the takeover of British Marconi interests in the United States by the newly formed Radio Corporation of America.[69]

The Federal Communications Act of 1934, which created the Federal Communications Commission, was also a product of the Roosevelt administration. The act provided, among other things, that licenses should not be issued to any radio stations owned by corporations with alien officers or directors, or a quarter of whose stock was foreign-owned. It has been suggested that this provision was specifically aimed against ITT.[70] By 1945 suspicion about the loyalties of transnational communications companies was so strong that the chairman of the Federal Communications Commission made a public statement to that effect.

On the other hand, while the debate over internationalism went on in the United States in the years between the wars, and business and government leaders remained profoundly split on the issue, ITT was viewed with a mixture of admiration and envy in European capitals. Sosthenes Behn was regarded as leading the vanguard of an expanding and highly successful American international capitalism, and he eloquently defended this view before the Commerce Committee of the House of Representatives.

The United States was still a year away from war when Henri Busignies arrived in New York, but all the confusing currents and trends of the previous twenty years, as well as fears about the immediate future, greatly increased the distrust of all foreigners. As a result of official suspicions the Frenchmen and their families were kept under surveillance for quite some time. Their movements were watched and their home telephones were bugged.[71] The army, particularly, reacted to them negatively.

Arrangements had been made not long after the first meetings in January for the ITT contingent to visit Colonel Colton at Fort Monmouth to discuss radar. When they arrived, Colton went directly to the point. "You are French, aren't you?" Deloraine recalls him asking. And then Colton added, "I wouldn't discuss radar with Jesus Christ himself." Understandably, after that, the meeting was very brief.[72]

Colonel Colton was to have a change of heart only months later, when he learned, to his chagrin, that the navy had gone ahead and given ITT development contracts. Colton then decided to give ITT the opportunity to develop radar equipment for the army, including pulse-transmitting tubes of ITT's own design. Eventually ITT supplied the Army Signal Corps with the SCR 291, 501, and 502 radar, portable D/Fs (a Busignies design), and a range of other equipment.[73]

The surrender of France to the Germans, and the subsequent collaboration of the Vichy regime, had added another element to the usual confusions and war-engendered suspicions that, at the very least, slowed down the assimilation of the Frenchmen into the American military effort. There was profound ambivalence surrounding the Vichy French— were they the legitimate government of all Frenchmen? And if so, what was the position of General Charles de Gaulle and his Free French organization?

On 31 August 1941, nine months after Busignies's arrival in the United States from Vichy France, the front-page headline of the New York *Herald Tribune* read, "Vichy Embassy in U.S. Shown As Heading Clique of Agents Aiding Nazis."[74] This was the fruit of the undercover work of "the Quiet Canadian," Sir William Stephenson, codenamed Intrepid. Officially Stephenson headed the British Security Coordination, with headquarters in Rockefeller Center in New York, but his actual assignment from the Secret Intelligence Service in London was to collect information on pro-German activities in the United States and to direct countermeasures.[75]

Stephenson's agents had infiltrated the Vichy embassy in Washington and had indeed discovered a nest of Nazi sympathizers. The Quiet Canadian quickly succeeded in arousing a profound nationwide distrust of Vichy France. He accomplished this by judicious "leaks" of information to the *Herald Tribune,* which were then reproduced in over one hundred newspapers in the United States and Canada. The resulting suspicion inevitably spilled over onto Busignies and his associates. This was especially irksome to them as they had undergone danger and hardship to get to North America in order to continue the fight against Nazi Germany.

It had been natural for Mme Busignies to play an active part against the invaders of her country. She had been raised on the tale of her great-grandparents' youngest son, who, in 1870, at the age of seventeen, had been executed by the advancing Prussians. "So you see," she explains, "Nazis in 1870 or in 1940, they are the same." Perhaps if her activities as a courier for the French underground had been made known to Washington, the reception of her husband and his colleagues might have been less cautious. However, Mme Busignies could not tell her own story because her father had taken her place working for the Resistance.[76]

So, probably for an aggregate of reasons, there was a pause in the development of HF/DF in the United States for a few crucial months in 1941, just as there had been a little earlier in Britain. And again security was an issue.

Among the records of the Bureau of Ships now lodged with the National Archives and Record Administration at their repository in Maryland, there is a box of files full of correspondence concerned with ITT and its subsidiaries (chiefly International Telephone and Radio Manufacturing Company and its successor, Federal Radio and Telephone Company). The bulk of the letters are dated between August 1941 and May 1943, and they are mostly on the topic of security clearances.

The ITT letters are from a variety of people but mostly from Sosthenes Behn himself. The letters are addressed to, among others, Mr. Ray Ellis of the Office of Production Management; to Col. R. B. Colton of the office of the chief signal officer of the army; to Dr. Jerome C. Hunsaker, the coordinator of research and development; to Rear Adm. Leigh Noyes, the chief of naval communications; to Rear Adm. Edward L. Cochrane, the chief of the Bureau of Ships; and to the Hon. Frank Knox, the secretary of the navy. Attached to them are replies from those gentlemen. In addition, the box contains a thirteen-page confidential "Case History" issued by the Office of Naval Intelligence (ONI) on 9 January

1943, on the subject of the "International Telephone and Telegraph Company and Subsidiaries, 67 Broad Street, New York, New York."[77]

Reading these letters shows clearly the importance attached to security issues in time of war, and the emotional nature of the debate over their application. The case history is a curious collection of innuendo, unfounded suspicion, false reasoning, malicious intent, and above all, sloppy investigation. And yet while serious charges were being considered, we know that in practice ITT kept contributing to the war effort. And its contributions were being gratefully accepted and used by the U.S. Navy and by the army.

On 14 August 1941 Sosthenes Behn opened the debate with a letter to Admiral Noyes. In it he refers to a visit about ten days earlier to Dr. Vannevar Bush, chairman of the Office of Scientific Research and Development. During that visit, wrote Colonel Behn, "I understood Dr. Bush to say that while he had previously received a clearance as to the International Telephone and Telegraph Corp. from the Navy Department as well as from the army, that some reservation had recently been made by the navy, and he told me that he proposed to make further inquiries with a view to clearing the situation."[78]

On Colonel Behn's second visit to Dr. Bush he learned that ITT's "limited" clearance, to which he objected, had not been changed. Colonel Behn did not understand why ITT could not "receive a general clearance from the army and navy giving it a status equal to other American companies." As he explained to Admiral Noyes:

> The International Telephone and Telegraph Corp. has a subsidiary manufacturing company, called the International Telephone and Radio Manufacturing Co., with factories at Newark and Laboratories in New York. The factories and Laboratories have received certain orders from the army and navy and it has always been understood that information in connection with such orders could not be passed on to the parent company (International Telephone and Telegraph Corp.) or to any of the foreign subsidiaries of the parent company. These conditions and limitations must undoubtedly exist for all American companies who had pre-war agreements for exchange of information with their subsidiaries or licensees outside of the United States. I believe all these facts are well known to both the army and navy but have repeated them so that everything should be made perfectly clear and well understood.
>
> May I therefore ask you, if there is any reservation or limitation on the part of the navy with respect to the International Telephone and Tele-

graph Corporation as an American company with the same rights and obligations as any other American company, that I be given an opportunity to submit to you any further information that may be needed, as you will readily understand it is important that we should know exactly where we stand with the National Services since the plant and Laboratories of our Newark subsidiary are now carrying out certain confidential orders and we are anxious to collaborate to the maximum in the interest of National Defense.[79]

Admiral Noyes forwarded the letter to the chief of BuShips as a matter under his cognizance. Colonel Behn's letter stated most of the major issues: that ITT had been denied a general clearance; that he believed this was on the specious grounds of ITT's pre-war international connections; that this rule was not being applied to other American companies with international affiliations; and that such a rule made no sense anyway, as some of ITT's subsidiaries were already engaged in secret government business.

But there was a further sore point that was soon to be broached. This was the suggestion that there was a suspicious (or at least questionable) "foreign" element to ITT's top engineering staff. In a letter of 14 August 1941, H. H. Buttner, ITT vice president, explained to Colonel Colton of the Army Signal Corps that ITT's capacity to engage in substantial war work was dependent on "bringing to America some British and French engineers whose reputations were outstanding in communication developments abroad, and also bringing to America technique and manufacturing information and drawings and in some cases models which had been developed abroad over a long period of years."

Buttner adds, possibly with a hint of irony (or was it exasperation?) that "Our associated company, Standard Telephones and Cables, Ltd., in London, has manufactured numerous items of military equipment for the British Services, equipment which has already proved itself in use." He ended by saying: "We have provided a well-qualified organization with factory facilities, and are prepared to do our part in producing promptly the quantities of radio equipment required for National Defense under the Emergency Act."[80]

So what was the problem? In May 1943, almost two years later, the issue of limited clearance was still outstanding. Secretary of the Navy Frank Knox explained to Sosthenes Behn, whom he addressed as "My dear Mr. Behn":

Because your company has in its employ certain alien engineers, and in addition, because it has substantial and widespread business interests in many foreign countries with which business intercourse must be maintained, it is not deemed advisable to offer full clearance. I believe you will recognize the necessity the navy is under in not offering full clearance where there is any possibility of leaks of highly classified information.[81]

Frank Knox compounded his topsy-turvy explanation by concluding:

It is desired to take this opportunity to state that the International Telephone and Telegraph Company and its subsidiary have made very substantial contributions to the war effort. The Department recognizes that the alien engineers in the employ of the company are of the highest caliber and that they have had a most important part in these contributions of the company and the Department desires to continue on the present satisfactory basis involving the full utilization of the company and its technical personnel whether they be nationals of the United States or France.[82]

On 6 May 1943 Colonel Behn responded. He wrote Secretary Knox that he had been prepared to accept the decision limiting ITT clearance on the basis of their employment of some alien engineers and technicians, even though they had been "associated with our Corporation for a great many years and for whose loyalty and devotion to our cause I have most completely vouched," providing this was a general rule with respect to navy classified contracts.[83] But Behn wrote that as far as he could tell, no such limited clearance had been imposed on any of the other large American companies that had comparable foreign investments. Behn "earnestly" requested Knox's reconsideration, submitting "that for us to be made an exception cannot but cast a doubt on our integrity and loyalty, with a depressing effect on the morale of our engineers and employees."[84]

Knox responded that he regretted he could not reconsider, even though there was no reason to doubt the integrity of ITT "in the light of any existing information."[85] Still, it is almost certainly because "information" about ITT had been collected that these clearance problems existed. In a memorandum (13 January 1943) from the director of naval intelligence (DNI) to the director of base maintenance division, on the subject of ITT and its subsidiaries, the DNI stated that "from time to time, information has been received . . . concerning general conditions in the subject corporations, and concerning more specifically certain personnel." As this information was received, the DNI passed it on to BuShips, to the coordinator of naval research, and to the director of naval communications. The memorandum went on to recommend that if any action was to be taken, it should be coordinated between the director of military intelligence, the chief of the Signal Corps, the director of naval communications, and the chief of BuShips.[86] This was pretty sweeping coverage.

Page one of the case history notes that BuShips had placed restrictions on the type of work being done by the International Telephone and Radio Manufacturing Company because of "information received" (not solicited, apparently) "indicating the undesirability of entrusting that company with confidential or secret information." What exactly was the nature of this disturbing information? It is set forth in detail in the next twelve pages of the case history.[87]

Under the heading "International Ramifications, Policy and Activities," the ONI cited the foreword of ITT's annual report from 1931, which indicated the international scope of the corporation. Then it quoted part of an address made by Colonel Behn, also in 1931, in which he said, "our headquarters and field staffs are open to all without preference or prejudice." Commenting on this, the report noted that by July 1941 ITT had subsidiaries in twenty-one countries, which, in the opinion of the navy member of the NDRC and the assistant coordinator of naval research, made it "so dependent for its very existence on the goodwill of Germany and other Axis powers which are in a position to control a large part of its assets, that we cannot risk information being transmitted abroad."[88]

The case history goes on to cite numerous other examples of the questionable foreign entanglements of ITT, and also to explain the source of some of its information. Typical of such entries is the following:

> Another informant, Comdr. George Shecklen, USNR, an official of the Radio Corporation of America (a firm competing with International Telephone & Telegraph) [in fact, the same RCA of competing D/F devices] stated that International Telephone & Telegraph has made a considerable effort to have certain priority equipment needed for the War effort shipped to South America for some of their subsidiaries.[89]

This whole issue of ITT's continuing business contacts with Axis powers during the war was revived in 1973 by a journalist, Anthony Sampson, in a book called *The Sovereign State of ITT*. Sampson repeated most of the rumors and charges about ITT that had been current during and immediately after the war.

One of the seemingly most damaging aspects of the Nazi connection was the purchase in 1938 by C. Lorenz, A.G., a German affiliate of ITT, of a 28-percent share in the Focke-Wulf company. Focke-Wulf was a German manufacturer of bombers that were used extensively during the war to bomb Allied convoys. Sampson maintained that Behn continued to control Lorenz during the war through contacts in neutral Spain and Switzerland. The South American charges involved the strong suspicion that ITT was allowing its telephone lines in Uruguay and Argentina to be used to pass on vital information about convoy sailings to German submarines.[90]

In 1982 a thorough, well-researched examination of ITT put Sampson's allegations in perspective. The author, Robert Sobel, concluded that "earlier accounts claiming ITT had actively cooperated with the Nazis were either incorrect or exaggerated and certainly hadn't drawn upon documents readily available at the National Archives."[91]

Professor Sobel shows that Behn was in constant contact with the State Department during the war, that he occasionally informed the State Department of developments at ITT's holdings in occupied Europe, and that from time to time he wrote directly to Secretary of State Cordell Hull to solicit his permission before taking some action in regard to these affiliates. Sobel also points out that the telephone companies in Uruguay and Argentina were government-controlled concessions. If they had failed to transmit any calls they would have been taken over by the undoubtedly pro-Nazi Argentine regime. In fact, Henri Busignies recorded that Behn contacted the State Department early in the war to see if they wanted him to try to impede the South American telephone communications with Europe, but he was told not to do anything. All the calls were already being monitored by Washington, which was listening with interest to what the German agents were telling Berlin![92]

Perhaps it is not surprising that the intricacies involved in maintaining an international business empire at a time of global war should have left ITT vulnerable to charges of disloyalty. Similar charges were leveled at many other major corporations, including Du Pont, General Motors, and General Electric. And it was well known that Ford Motor and Standard Oil of New Jersey maintained some sort of relations with the powerful German conglomerate I.G. Farben, at least during the early stages of the war.[93]

Apparently what was most damning for ITT was the foreign appearance of Sosthenes Behn himself, which tinged his whole corporation.
Years later Deloraine recorded with amusement that while the Colonel
seemed "a typical American" to Europeans, to Americans he was dis-
tinctly different. As a cosmopolitan American executive once explained
to Deloraine, Colonel Behn "does not dress like an American, he does
not eat like an American, and he carries his handkerchief in his sleeve—
which no American does."[94] Behn was also suspected by some of being
a convert from Judaism (though there is no evidence for this and Behn
was actually a devout Catholic), and by others of being a Nazi sympa-
thizer, a personal friend of Hermann Göring. The rumors surrounding
this exotic, mysterious man who never granted interviews must have
fueled the already overheated imaginations of many security investiga-
tors. Even the name Sosthenes (Greek for "life strength") was clearly an
oddity.[95]

Section three of the case history, which lists "General Activities of a
Suspicious Character," expresses this chauvinism quite transparently.
The first item addressed is HF/DF, which is mistakenly identified as radar:
"Before the war, French scientists and telephonic experts connected with
the French subsidiary of International Telephone & Telegraph, were working on radar equipment to be installed on warships, for the detection
of submarines. This equipment was supposed to have been destroyed
before the German invasion, but was not."[96]

Apparently the knowledge that this equipment was already being used
by the U.S. Navy in its shore-based HF/DF network was not available
to the ONI.

The rest of the document deals mainly with personnel. One ITT vice
president, Trevor H. Clark, was denied clearance and access to the Naval
Research Laboratory because while in Lisbon in January 1941, awaiting
passage home to the United States, he "dined with one Wenzel, a well-
known German agent and four German girls."[97] With regard to Colonel
Behn, the report acknowledges that he had served as a communications
officer in the U.S. Army during World War I, but points out his "difficult position as an international industrialist and promoter."[98] This re-
flects the same perversity that employed young Japanese-Americans in
the U.S. Army while their "suspect" families languished in concentration camps.

Also investigated was an employee identified only as "La Blanc," who was apparently "a rabid isolationist, a member of the America First Committee and the Christian Front, and an ardent follower of Father Coughlin." On the other hand one Walter Ryan, a cable operator at ITT, was reported to have "talked radically since the United States entered the war." The implication here was that Ryan had left-wing or communist sympathies. Of course, at that time the United States and the USSR were allies, but that did not alter the fact that during the war the definition of a politically sound American was a very narrow one. It is no surprise to find that the source of most of the information in the case history was from confidential "reports" made by unnamed but "reliable informants."[99]

Perhaps most illustrative of the general confusion of the situation is the following statement about Deloraine:

> On August 16, 1941, the Chief of Naval Operations advised there was no objection to the employment of Maurice Edmond Deloraine. However, on August 29, 1941, information was received from the Military Intelligence Division which indicated there was a reasonable doubt as to the desirability of the Chief of Naval Operations recommending such consent, and on September 22, 1941, the Federal Bureau of Investigation was requested to investigate Deloraine. [Perhaps in response to the newspaper exposé of the Vichy diplomats?] On December 1, 1941, the Secretary of War revoked his temporary consent. Finally, on January 8, 1942, the Secretary of the Navy consented to the employment of Deloraine "provided that his activities be confined to the work" on which he was engaged as of the date of the consent (direction finding equipment).

Apparently, one of the most troubling details in Deloraine's dossier was that he had secured his and his colleagues' escape from France by "bribery of German officials."[100]

This same roller-coaster pattern seems to have prevailed with regard to security clearances for all four of the French engineers. Finally, between January and March of 1942, the secretary of the navy definitively consented to the employment of all of them. Maybe coincidentally, on 26 March 1942 Deloraine and Busignies were invited to the navy conference where the shipboard HF/DF program finally got under way.

To put this issue into perspective it may be helpful to read part of the section on "Security" in the Naval Research Laboratory's war history, bearing in mind that it was written in 1946.

Scientists [at the NRL] frequently broke the security regulations by retaining confidential or secret documents, not through carelessness, but because it often took two weeks or more for a document to pass through the overburdened filing mill, and in some cases reports were mis-filed and could not be located until long after the need for them had passed.

In spite of all the safeguards, the Laboratory probably would have seemed rather insecure to a trained intelligence officer. Anyone with a plausible excuse could wander at will through the corridors; highly classified material stood for weeks in rooms open in the daytime; and classified documents were left unguarded on desks and laboratory benches. (The Security Officer has stated that the worst offenders were the higher ranking naval officers and civilians.) Civilian scientists have remarked that the secrecy regulations imposed by industrial plants were far more restrictive than were those of the navy's own laboratory; shop employees have stated that it was easier to get into or out of the station during the war than it was before or after."[101]

ITT's own security regulations compared very favorably with those of the NRL. A letter of 22 January 1942 from the International Telephone and Radio Laboratory to the director, BuShips, explained the "special precautions," which had been taken to "insure the security of information and secrecy of documents entrusted to the Lab. In the first place, all employees are fingerprinted and investigated by the FBI and are reported to corresponding bureaus of the army and navy upon employment." Special ID badges were worn at all times, and the premises at 67 Broad Street and the other labs were under armed guard twenty-four hours a day, seven days a week. Visitors were only permitted to enter the labs upon receipt of special authorization from U.S. authorities, and all visits to the labs were reported monthly to the services. All classified material was kept in special safes.[102]

Nowhere does the case history explain why, in May 1943, Sosthenes Behn was still trying to obtain a general security clearance for his corporation, when its "suspect" employees had been cleared and when it was heavily engaged in classified defense work. Apparently those who made the decisions with regard to what technology was required for the prosecution of the war were able to secure that technology in spite of the qualms of the intelligence establishment.

After the spring 1941 meeting in Washington at which high-ranking navy officials approved the testing of Busignies's shore-based HF/DF, the next record of contact between the navy and ITT on the subject of direction finders dates from the fall. On 10 October the chief of naval operations (CNO) requested an additional fifteen DAJ-type HF/DF sets to be delivered to BuShips for installation at specified sites. The CNO's request stated: "We believe that the Busignies direction finder is the best suited to fulfill the strategic direction finder requirements of the navy."[103] This was certainly a very strong endorsement of ITT and of Busignies's high-frequency direction finder.

The original group of four Frenchmen, working at labs in Great River and in New York City, had grown to about 150 people by then, mostly engineers and designers. At this time two American engineers, Avery G. Richardson and Eugene Lombardi, joined the group. The production of the first four DAJ devices for the experimental model had been handled easily by subcontracting most parts and insisting on rapid delivery. But with the next order a new problem arose.

The order called for fifteen complete sets of equipment, each set involving four direction finders, in order to maximize the number of frequencies that could be "guarded," or watched, at one time. Federal Radio and Telephone Company (which in 1942 became part of the International Telephone and Radio Manufacturing Corporation) was at that time the only ITT manufacturing company in the East. It was very small, employing perhaps two hundred people, and it could not supply the expanded manufacturing needs brought on by military contracts. Most parts were therefore obtained from subcontractors, and then assembled and tested at Great River and other new sites rented for this purpose.[104]

Deloraine noted in his account of the development of Huff Duff that after Pearl Harbor the atmosphere in the United States quickly changed, and previous suspicions of the Frenchmen seemed to have disappeared. "The navy gave us every help to speed up our program and our relations with them became very good," he remembered.[105] It seems the intervening thirty years, or goodwill, had dimmed his recollection of the annoyances occasioned by naval intelligence. Perhaps, too, they were largely ignored at the time.

But even before the United States was actually at war, a new question had finally been raised by the navy that was to be of great significance to ITT, especially to Busignies. This was whether or not HF/DF

was being used to maximum advantage by depending on shore-based installations. The CNO praised the strategic importance of Busignies's HF/DF device, but what of tactical use? The British had shown that HF/DF equipment could also be developed for installation on ships, though with some loss of range and accuracy due to the limitations in antenna size.

In memos of 31 July and 15 September 1941, Admiral King (then commander, Atlantic Fleet, and still in his initial offensive stage) had drawn the attention of the CNO to the use of this equipment in foreign navies, recommending its further development in the United States. As a result, on 1 October the CNO's office advised BuShips to procure fifty portable sets of a type proposed by RCA, or other similar models. On 14 October the navy further requested that BuShips obtain three or four sets of each type of British shipboard direction finder in order to assist the American development program.[106]

It was to be five more months before any decisive action was taken. On 26 March 1942, a BuShips-NRL conference on the development of U.S. shipborne HF/DF included only two non-navy representatives: Deloraine and Busignies of ITT.[107] How this ITT coup came about is unclear, but as usual it was probably the result of Colonel Behn's great energy and extensive contacts. Industry once more gave the navy a nudge in the right direction, though there is no further mention of the competitive HF/DF devices from RCA.

When the Frenchmen arrived in Washington for the second time, they found, as before, that there were a number of senior officers gathered around a table. But this time there were also models of ships' superstructures on the table. After being questioned about their ability to deliver a shipborne HF/DF, the engineers were asked where they intended to place the antenna. Busignies pointed to the top of the highest mast and asked if that was the radar antenna. The answer was yes. "Then I propose you remove it," Deloraine quotes Busignies as saying, "and place our antenna there instead."[108]

Tech. Sgt. Eugene J. Lombardi (*right*), in London in 1942 to check installation of SCR-291 portable D/Fs for U.S. Army Signal Corps.
Courtesy of Eugene Lombardi

Deloraine subsequently recorded that he thought this bit of audacity would get them thrown out of the room. On the contrary, in this encounter between radar and HF/DF, the latter held its own. It was decided that representatives

from the NRL and ITT's International Telephone and Radio Laboratory would test the shipboard HF/DF model—DAQ—being developed by Busignies. They were to compare its performance at sea with the British type-FH3 equipment aboard the destroyer *Corry*.[109]

The *Corry* tests were set up in May by Maxwell Goldstein. He concluded that the FH3 and the ITT models gave "as may be expected . . . somewhat equivalent performances," and BuShips recommended the purchase of the ITT equipment.[110] On 26 June 1942, the navy approved a CNO proposal in which half of all new-construction destroyers destined for the Atlantic Fleet and half of all destroyer escorts (most of which were for Atlantic service) would be equipped with HF/DF of ITT design.[111] This signified a recognition, obtained from recent more encouraging British experience, that HF/DF could be at least as important as radar to the anti-U-boat campaign.

At a similar meeting with the navy some time later, Maxwell Goldstein also recommended, with all of Busignies's temerity, that HF/DF replace radar on the masthead of half of all escort vessels.[112] Happily, it was eventually found possible for the same ship to carry both HF/DF and radar antenna, though generally on separate masts, and they were deployed most effectively together.

On 26 March 1942 the U.S. Navy had decided to design and build a shipborne HF/DF, and by May the equipment was delivered, installed, and under tests on board the *Corry*. After the successful *Corry* tests, everything was organized at the various ITT companies to produce this vital equipment at top speed, while still maintaining quality. To meet these demands, ITT and its many subsidiaries and associated companies grew at a tremendous pace. This was especially notable in the research and engineering division—the International Telephone and Radio Laboratory (later the Federal Telecommunications Laboratories). As the armed forces became increasingly aware of the high quality of design and engineering that had by now become a trademark of the Busignies group, they requested studies of numerous new projects involving various types of D/F for land, ships, submarines, and aircraft.

Destroyer *Corry* at Annapolis, May 1942, showing FH3 HF/DF antenna loops aft.
Naval Research Laboratory

During the war, ITT laboratories developed and supplied, directly or via other companies, thirty-three different types of equip-

Corry, May 1942. Port side looking aft shows relationship of FH3 HF/DF antenna loops with stacks,
superstructure, and rigging.
Naval Research Laboratory

Corry, May 1942. According to NRL Report No. 1896, this shows "ITR&L [should be IT&RL, or International Telephone and Radio Lab] loops on foremast." In fact, no HF/DF loops are visible.
Naval Research Laboratory

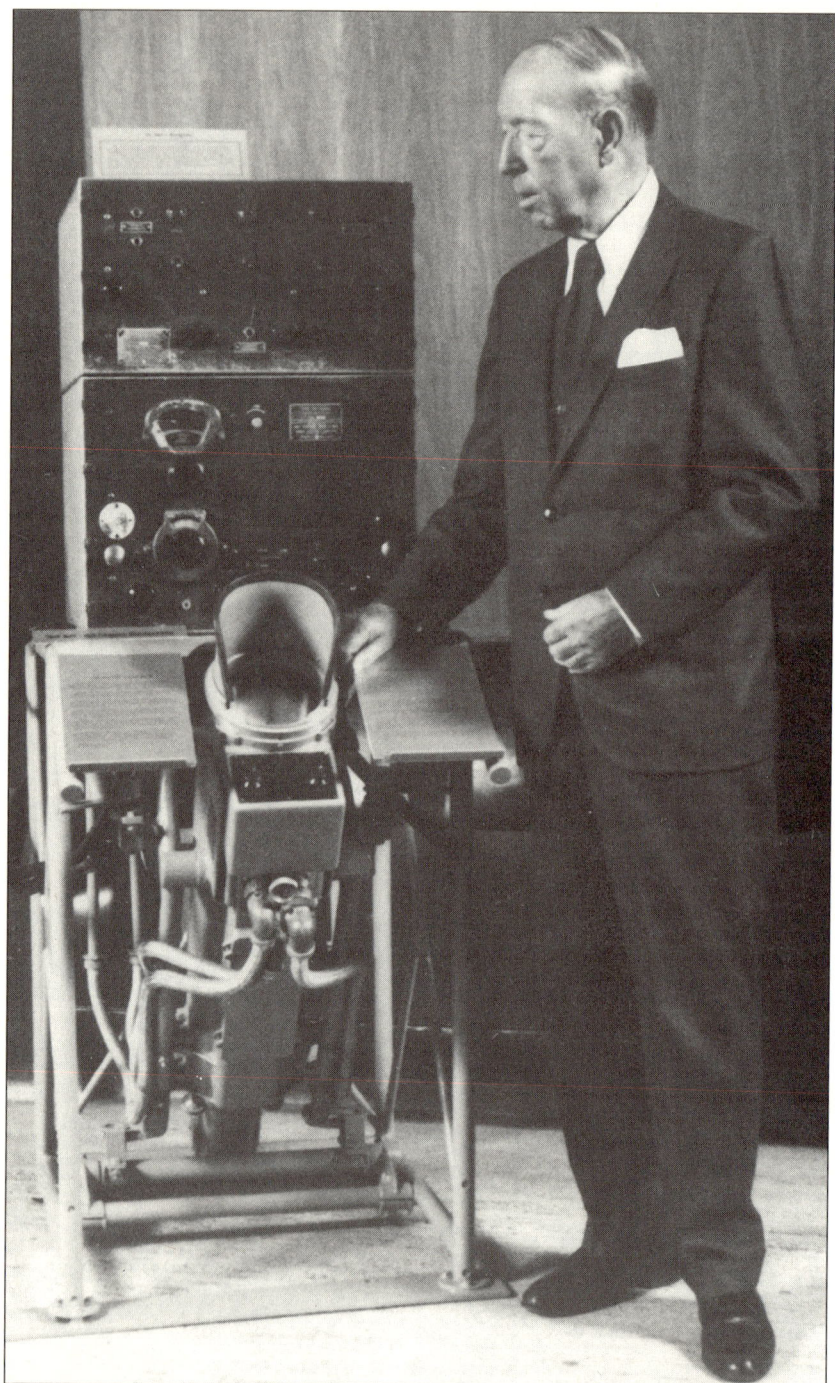

ment, some admittedly in small quantity. All told, they delivered approximately four thousand direction-finder sets. These successes grew out of the work that ITT had been doing in Paris since 1930. By the time hostilities ceased, the laboratories had grown from the original four Frenchmen to over 950 scientists, engineers, technicians, and administrative personnel. Facilities at Great River and in New York City were gradually expanded to cope with wartime demands, until in 1945 the first unit of a modern laboratory at Nutley, New Jersey, was opened. That site now houses the ITT aerospace labs. [113]

Between 1941 and 1945, ITT received over 160 development contracts from the U.S. Army, Navy, and National Defense Research Committee. These contracts, mostly of a secret or confidential nature, varied in scope and included vacuum tubes, radar, communications, aerial navigation, and electronic countermeasures equipment.[114] Still, it was direction finding that was the most important contribution of the company to the Allied war effort.

An article published in 1946 in ITT's technical journal *Electrical Communication* claimed that Federal Telecommunications Laboratories were "the only supplier of automatic direction finders to the U.S. Navy. Thousands of these equipments were installed on ships of all sizes and types."[115] Nor did research end with the development of the DAJ land-based and DAQ shipboard models. By February 1943 there were the DAK intermediate frequency, visual D/F, and the DAG portable D/F. Later in 1943 a new model, the DAU, came into use. This was an improved version of the DAQ for ships, and it featured a new panoramic scanner that increased the speed and accuracy of bearing indication. It was followed in 1944 by the DAW truck-mounted model for mobile use on land, and the DBA for ship use with a small, continuously rotating, resonant, shielded loop.[116]

In recognition of these contributions, in March 1943 the laboratories were awarded the army-navy E for excellence in production. This was followed in 1944 and 1945 by three stars to be placed on the E pennant as a sign of continued excellence. Sosthenes Behn was awarded the Medal for Merit, the highest civilian decoration for wartime services. Following the end of the war, public releases of the

Henri Busignies in 1978 with his WWII shipborne DAQ HF/DF operator's station.
Courtesy of Cécile Busignies

army, navy, and NDRC paid great tribute to the part played by the laboratories in the war effort. Much of the credit for this was due to the brilliant contributions of the French engineers who risked their lives and those of their families to escape from occupied France and continue their anti-Nazi campaign from the United States.[117]

In 1945 Dr. Deloraine was made an officer of the Legion of Honor by the Minister of the French navy for his part in this work. He also received a Certificate of Commendation from the U.S. Navy and a Certificate of Appreciation from the Office of Scientific Research and Development. Dr. Busignies was awarded the Certificate of Commendation for "outstanding service" to the navy, and the Presidential Certificate of Merit "for his outstanding contribution to the war effort of the United States and its allies." Yet for a time in 1941 it had not been at all clear that the service these men offered would be acceptable in the country to which they had fled.[118]

CHAPTER 6

The Decision Makers: Individuals and Organizations

I
n the national struggle for victory, no source of ideas for the improvement of naval materiel is neglected." That, at any rate, was the claim of Rear Adm. A. H. Van Keuren, then director of the Naval Research Laboratory, in a March 1944 article in the *Journal of Applied Physics.*[1]

It had been almost a year since the tide had turned against the German U-boats in the Atlantic, and it was fairly clear that victory was just a matter of time. Part, at least, of that success was due to the contribution of effective land-based and shipborne HF/DF, the American versions of which were the product of Henri Busignies and the ITT laboratories. Yet at the same time that ITT's foreign engineers had been working so industriously for the United States, in some quarters they had been under suspicion of posing a security risk.

If Admiral Van Keuren's view represented the pragmatism of a scientist who just wanted to get the job done with whatever tools were at hand, as his comment suggests, then who was it who had opposed the French engineers? How much had that opposition retarded the development of HF/DF in the United States? The answer to these questions may emerge if it can be found where the decision-making power really lay in the complicated structure that was the U.S. Navy at war.

What kind of influence was wielded by the president, the secretary of the navy, the top navy officers? What influence might strategic decisions have on the development of specific vessel types, weapons, and technical devices, and how far down the chain of command did the leaders reach to make their influence felt? At lower levels, what institutions, groups, and organizations, both civilian and military, were involved in research and development of scientific and technical devices for the navy?

How did those groups interact with each other and with the strategic decision makers? Did private industry have any influence on naval decisions? In other words, how did the U.S. Navy go about getting what it needed to fight the Atlantic campaign, and, more specifically, who generated the ideas about naval electronic requirements and who decided if those requirements should be met and how?

As the commander in chief of the armed forces, the president of the United States had the authority to make final decisions on military matters. As President Roosevelt was particularly interested in the navy, he played a significant part in the debate over what shape the naval war was to take. Even before the United States was actually at war with Germany, Roosevelt deployed the navy actively in the Atlantic. At first in the guise of the so-called Neutrality Patrol, and then on actual convoy escort duty in concert with British and Canadian naval forces, American ships were involved in thwarting German submarine attacks on merchant vessels. In November 1940 Roosevelt had agreed with Churchill that no matter what else happened, Germany was to be defeated "first." After Pearl Harbor was quickly followed by a German declaration of war, the U.S. Navy found itself fighting on two oceans at once. But still Roosevelt remained a keen advocate of dealing with Germany first, and only then of turning to Japan.[2]

Convoy-escort duty in the harsh weather conditions of the North Atlantic, however, did not have the same appeal as the clash of mighty battle fleets in the Pacific. Career officers were afraid there would be little chance to advance in their profession while playing sheepdog to slow, ungainly, unglamorous merchant ships. There was also an institutionalized predisposition in the professional navy to see the Pacific as the major forum for a modern naval contest. So Roosevelt struggled to keep the focus on the Atlantic and to secure for the Atlantic campaign the close attention he believed it needed. And in this, as in most naval issues, Roosevelt's intervention was decisive.[3]

To be sure, the president's interest was sometimes a mixed blessing. Roosevelt exerted pressure not only on major strategic issues but also at much lower levels, down to the promotion of specific officers and the types of vessels that should be produced. Some of the developments backed by Roosevelt proved to be successful, like the vital construction of escort carriers instead of only large fleet carriers, which Admiral King wanted. But other issues, such as his preference for small, relatively ineffectual submarine chasers, actually inhibited the production of the larger destroyers and destroyer escorts that were better suited to the rigors of antisubmarine warfare. On balance, though, Roosevelt's role as decision maker was a good thing for the Atlantic campaign. His interest in and support of the Atlantic Fleet encouraged the navy to take action in spite of some internal doubts as to the significance of that theater.[4]

Secretary of the Navy Frank Knox had the political power to push Roosevelt's wishes on the navy even if it was reluctant. In fact, however, War Secretary Henry Stimson appeared to be far more concerned than Knox about the navy's role in coming to grips with the U-boats. Like ITT's Sosthenes Behn, Stimson relished being called by his former military title, Colonel. The two men had even more in common—Stimson, who "believed strongly in the offensive," was also an enthusiastic supporter of scientific developments that could improve war-making capacity. Stimson openly disparaged Knox's lack of control over his own department, and Stimson had a point.[5] Knox's influence on the navy between 1940 and 1944 seems relatively minor in comparison with the influence of the exceptionally able, aggressive, and determined admiral to whom Roosevelt gave unprecedented power on the outbreak of war.

On 30 December 1941 the president selected Adm. Ernest J. King to become the new commander in chief, U.S. Fleet. In March 1942 King also assumed the position of chief of naval operations, combining control of the navy bureaus with operational command of all fleet units. A year before Dönitz took over leadership of the German navy, King was, in effect, given command of the entire U.S. Navy, which gave him exceptional authority.[6]

Arrogant, irascible, and domineering, Ernie King certainly set his stamp on the navy. Although few liked him, King was a brilliant administrator and an exceptionally able sea officer. Above all he was tough, and that was what Roosevelt wanted at this time of crisis. Nevertheless King's appointment was curious and, many have argued, counterproductive, at least for the Atlantic campaign. It was well known that King's focus of interest was in the Pacific, and also that his virulent Anglophobia extended forcefully to include the entire Royal Navy.[7] King's antipathies were not to be taken lightly; his foul temper was legendary.

It has also been claimed that King never really accepted the validity of "Germany first." But since he was in command of the Atlantic Fleet when that doctrine was propounded, and since he himself whipped it into effective shape when he took over in December 1940, the case against King is clearly overstated. After all, it was his dynamic leadership of the Atlantic Fleet that impressed Roosevelt so much that it led to King's subsequent promotion. But to whatever degree King's view was biased, especially with regard to Britain, this was certainly an inhibiting factor in waging an effective Atlantic campaign.[8]

Moreover, the war against German U-boats was highly technical, and technology was King's great weakness. King has been called a "superb strategic thinker," but the same commentator adds that he was "essentially indifferent to those aspects of his responsibilities not directly related to naval operations and fighting."[9]

It was not until October 1942, therefore, that the commander in chief, U.S. Fleet began to take the necessary serious interest in ASW, with the institution of bimonthly ASW conferences. Although King had an evident bias against civilian scientists, their ideas gradually began to take hold.[10] In a curious twist, Dönitz, who may not have had any prejudice against the advice of technical experts, nevertheless failed to make effective use of independent scientific resources; in King's navy excellent use was made of operations research, in spite of the admiral's well-known scorn.

However strong King's personality and influence might have been, the size and dispersion of the forces he controlled made it impossible for him to achieve the same degree of personal domination as that exercised by Dönitz over his U-boat force. Nevertheless, until 1943 King's lack of enthusiasm for ASW must at least have blunted the U.S. Navy's effort against Admiral Dönitz's submarines.

Roosevelt was fully aware of King's relative indifference to the Atlantic, and he kept a close eye on events, intervening personally to ensure that the American part in that operation would not be neglected. It should be noted, though, that King was not alone in many of his judgments about Atlantic strategy. In the spring of 1943, when Allied shipping losses were at their highest and few convoys were reaching Britain unscathed, it was the Joint Chiefs who refused to divert vessels from North African troop transports to supply British domestic needs. Roosevelt had to overrule the Joint Chiefs and force them to cut back their demands for military shipping so that the commitment to provide vessels for convoys to Britain could be met.[11]

King's efforts to keep all power centralized in his office initially caused problems that were quickly blamed for a supply bottleneck in the Pacific as well as in the Atlantic. In spite of his great administrative gifts, King's insistence that the ultimate decision on all equipment should pass through him, certainly slowed the supply system. However, there is also plenty of evidence that before King took over, there was not really any system at all. In August 1941 the president had approved the construction of escort vessels for U.S. naval use. It took Adm. Harold Stark and his planners until February 1942 to come to a decision on the number and types of ships required. In this case construction had been delayed almost six months by bureaucratic inertia.[12]

That same February, a month before King became CNO and after a year of waiting, Adm. Royal E. Ingersoll addressed a request to the chief of naval operations to expedite the delivery of "radio-receivers with loud speakers for use at Conning Stations" of all vessels of the fleet. Ingersoll reminded the CNO that he had already "indicated general approval of these proposals and stated that the procurement of the necessary material was in hand."[13] Ingersoll's request had then been passed on by the CNO to BuShips. But there, apparently, the request had stalled. Trying to be helpful, Ingersoll pointed out that for some of the vessels, receivers of a lower standard of security and reliability would do, "if shortages must be considered." Finally he pleaded: "Any inexpensive receiver available in the commercial market will provide for these types the solution of this problem which has been a matter of concern to the commander in chief, Atlantic Fleet, since he first requested equipment for this purpose in January 1941."[14]

More than a year had gone by, and Ingersoll was still awaiting action on his radio receivers. Surely King's centralized control did not make matters worse.

The decision was made to proceed with the development of shipborne HF/DF in March 1942, the same month King became CNO. Yet delays occurred in the introduction of HF/DF too. Some of these delays were caused in the course of implementation, which will be examined in the next chapter, but the rest resulted from continuing confusion in supply.

The Bureau of Ships was theoretically responsible for the procurement of equipment for the navy, excluding weapons—which came under the provenance of the Bureau of Ordnance—and planes and related aviation equipment, which came through the Bureau of Aeronautics. At the beginning of the war, BuShips was handicapped in terms of decision making in the field of electronics, because no procedure for assessing electronic requirements had yet been established. So in March 1942 (just when the decision on shipboard HF/DF was being made), BuShips began a thorough study of the physical and electrical characteristics of each ship type. The study was completed by summer, and the Bureau was ready to make recommendations to the CNO with regard to radar and other electronic needs.

But the initiative for actually ordering electronic equipment rested with the CNO. He had the authority to establish allowances for ships and shore stations, and all other orders for specific equipment came to BuShips through his office.[15]

Still, eager as King was to maintain his authority in all fields, he did not actually offer much guidance with procurement. For all his ability and vision, King "was impatient with and remarkably innocent of the daunting complexities of meshing navy material requirements with American industry for the greatest naval buildup in history."[16] Had King shown more interest, the buildup might have proceeded more expeditiously.

In defense of Admiral King, it is only fair to point out that he was faced with a desperate situation in the Pacific in the early stages of the war. It is a mistake to underrate the enormity of his tasks as the Japanese made one conquest after another following Pearl Harbor, and it is hardly surprising that he was preoccupied with the Pacific theater. Even Stimson, who seemed so concerned with the Atlantic, did not rush to make the army's long-range B-24 bombers available to the navy for use in Atlantic ASW patrols. Just a few of those bombers would, and eventually did, make a great difference against the U-boats.[17]

King's reputed lack of interest in the Atlantic became relevant to the development of HF/DF when the question of strategy against U-boats was addressed. Once war was official, Admiral King's prime objective in the Atlantic was to protect military troop and war-materiel convoys to Europe. Never an opponent of convoy, King was just unwilling to divert the few ships he had available in those early days from their primary task of escort of military forces to the secondary goal of protecting mercantile coastal traffic. He also continued to believe, as did most of his advisers, that inadequately protected convoys were in greater danger than vessels sailing independently, especially when the merchantmen could be individually routed close inshore.[18] This was an excusable error, but then King prolonged and compounded it by refusing to accept British experience and advice to the contrary.

Finally, the disastrous U-boat attacks on American coastal traffic early in 1942 obliged King to organize adequate defensive measures to protect shipping in U.S. and Caribbean waters. He was forced to this step even if it meant pulling back from commitments to transatlantic convoying and even from commitments in the Pacific. Perhaps as a consequence of the drubbing he took over losses off the East Coast, he abandoned the idea of offensive ASW patrols. He now supported a conservative, defensive, convoy-protection strategy for all Atlantic traffic. Shipborne HF/DF could be seen as a still-experimental offensive and tactical weapon, one whose success was unproven and whose very effectiveness was seriously questioned. Such a device was unlikely to have elicited enthusiasm from Admiral King, especially in this defensive phase.

When the belated institution of convoys along the East, Caribbean, and Gulf Coasts in the spring and summer immediately and dramatically reduced sinkings, Dönitz moved his main U-boat effort back to the North Atlantic. There, in the absence of Ultra during the prolonged blackout of 1942, Allied success in routing convoys around U-boat concentrations was very limited. The growing number of U-boats deployed in ever-longer patrol lines also made it increasingly difficult to avoid them. While they faced the potent wolf packs, the major preoccupation of the Allied escort forces—British, Canadian and American—was defense of the convoys in their care. But as soon as some British escorts were fitted with HF/DF, they once again began to evolve more aggressive tactics.

Although the new offensive hunter-killer doctrine against U-boats was announced in at least one U.S. naval district as early as April 1942, this strategy did not really come into its own in the U.S. Navy until May 1943, when the Tenth Fleet was organized to oversee and direct the anti-submarine campaign. This change was due in part to the pressure brought to bear on Admiral King by the Combined (Allied) Staff Planners, who issued a report on 1 March 1943 on the subject of "Measures for Combatting the Submarine Menace." The report stated in part: "Defensive measures alone, however, will probably be ineffective unless supported by sufficient additional forces to permit offensive action."[19]

By May 1943 DAQ-equipped escorts began to be deployed in increasing numbers and became part of a multifaceted offensive campaign. Sustained American attacks on U-boats, resulting in sinkings, date from that time onward. To accomplish this goal had required an offensive strategy and aggressive tactics using escort carriers with their air power, more and better escorts armed with better weapons, and improved training. To achieve a good rate of sinking also required detection devices of superior performance, such as the new sonar models, centimetric radar, and shipborne HF/DF; it also required effective operational use of Ultra intelligence.

Admiral King had had to be pressured to agree to this aggressive ASW campaign, and, except for the fact that the decision to develop shipborne HF/DF was made on his watch (though just barely), he can take little credit either for the change in doctrine or for any of the subsequent successes. Fortunately there were those in the naval command structure who were less reluctant to exploit every resource in the antisubmarine war. Admiral Ingersoll, who on 1 January 1942 relieved King as commander in chief, Atlantic Fleet, was fully aware of what needed to be done.

Promptly—in early February—he formed the Antisubmarine Warfare Unit in Boston, under Captain Wilder D. Baker, formerly a destroyer squadron commander with considerable practical experience. Baker's task was to study antisubmarine procedures and supervise training in their application, to encourage the development of new weapons and devices, and to coordinate the dissemination of up-to-date information. Baker eventually became the U.S. Navy's chief expert in ASW.[20]

Ingersoll's Atlantic Fleet was to bear the major American share of the fight against Dönitz's U-boats. By establishing a special unit under his command to teach effective ASW techniques, Ingersoll exercised some influence on decisions as to equipment, tactics, and training to be used.[21] But doctrine properly evolves out of experience, and the U.S. Navy in early 1942 had had little actual contact with U-boats, since few of the convoys it escorted in the last six months had been attacked.

Perhaps aware of this, Ingersoll, no Anglophobe, sent Captain Baker to England in March 1942 for briefings on British progress in antisubmarine warfare. The Royal Navy's Operational Intelligence Centre was already providing highly effective and centralized control of Britain's war against the U-boats. Acting as a collection point or central clearing house for ASW information, the OIC received intelligence from all sources, details of technical advances, and the results of operational experience. It then analyzed these data and relayed its findings directly to the operational commands concerned.[22]

In April, on his return from England, Captain Baker organized a group of operations analysts, "Baker's Dozen," under the leadership of Philip M. Morse of MIT.[23] His hope was to initiate some sort of organization of the American anti-U-boat effort along the lines of what he had seen at OIC. In June Baker's Dozen became ASWORG (Anti-Submarine Warfare Operational Research Group, later Operations Evaluation Group) and was moved to Washington, where it reported directly to the commander in chief, U.S. Fleet. ASWORG (Group M under the commander's control) was composed of notable academics and scientists; it conducted research and analysis, particularly from studies of action reports. It produced the most penetrating studies of ASW then available to the U.S. Navy.[24]

Among the reports from Captain Baker's group there are a number on the subject of shipborne HF/DF, but these mostly concern the operational use of an already existing model. The reports were produced by Comdr. Paul R. Heineman. He became enthusiastic about HF/DF at sea after his experience with convoy ONS 102 in June 1942, when he first saw it in action on the Canadian destroyer *Restigouche*. Commander Heineman, leading Escort Group A3, served alternately on the two Coast Guard cutters that were the first U.S. vessels to be fitted with HF/DF, the *Campbell* and the *Spencer*. It was a converted British type FH3, the DAR, which both ships had acquired by November 1942. Commander Heineman's wholehearted endorsement of shipborne HF/DF, while not the reason for its original acceptance, was clearly influential in encouraging its further development and deployment.[25]

In general, Ingersoll's efforts to approach ASW methodically went ahead without much encouragement from Washington, and it was almost a year before a central organizational structure anything like the OIC was created. As mentioned before, it was not until May 1943 that King finally agreed to establish an administrative entity, the Tenth Fleet, to fulfill the same functions in the U.S. Navy that the OIC fulfilled for the Admiralty. Based on the British pattern, Tenth Fleet, under the very capable Rear Adm. Francis S. "Frog" Low, finally assumed overall control of all aspects of ASW.[26]

It might with reason be assumed that most of the equipment that the U.S. Navy used in the Atlantic was produced as the result of requests for particular devices and technologies coming direct from Admiral Ingersoll's fleet. But in fact, input from the operational forces was only one factor in the process of deciding what materiel the U.S. Navy would develop. Suggestions might also be initiated by any one of a large number of entities including research units such as ASWORG, navy bureaus such as BuShips, and organizational structures like the Tenth Fleet. There were also a large number of interservice and civilian groups, each of which could and did play some part in deciding what the U.S. Navy needed most to prosecute the war.

Once needs had been identified, there were many convoluted and constantly changing paths that might be followed to obtain the desired equipment. The crisscross of several interwoven layers of authority led to damaging delays in getting the navy what it required. But ultimately, out of the confusion emerged the vessels, weapons, and devices that made possible the U.S. contribution to the victory in the Atlantic.

The Naval Research Laboratory was the major U.S. naval research facility, and it was the body that had the most influence on the development of U.S. naval HF/DF. The NRL of World War II was in many ways the brainchild of two Edisons, father and son. As early as 1910, Thomas A. Edison had recognized the need for a research organization within the navy. Edison had succeeded in interesting Josephus Daniels, then secretary of the navy, in the project. As a direct result of Edison's suggestion, Congress in 1916 appropriated the money to establish an experimental and research laboratory. Naval Appropriation Bill 1916 H.R. 15947 stated, in part, that the funds approved were "for laboratory and research work . . . for the benefit of the Government service, including . . . the employment of scientific civilian assistants as may become necessary."[27]

Because of other more pressing needs on American entry into World War I, the opening of the laboratory was delayed until 1923. It began its operations in a small group of buildings on the Potomac River in Anacostia, Virginia, across from Washington, D.C., where it is still located today. Originally composed of two main divisions, the Radio Division (which was to conduct the tests on radar and HF/DF during World War II) and the Sound Division (where sonar research took place), the laboratory kept adding other divisions as it took on an ever-widening scope of research in succeeding years. Close ties with outside scientific establishments were built into the original structure of the NRL; after all, the inspiration for the laboratory had come from civilians. Indeed, civilian scientific and military cooperation was functioning at the NRL from the beginning.[28]

The laboratory had initially been founded chiefly to conduct pure research. But in the late 1930s, as international tensions grew, the need was felt for practical devices ready for immediate application and the focus shifted more and more to applied research. Most theoretical studies eventually became an unaffordable luxury, and in wartime the NRL was predominantly occupied with development and test work. Most of this work continued to be undertaken, as it had been earlier, in response to specific requests from navy bureaus.[29]

The Radio Division of the NRL anticipated the prewar push for many practical devices. In 1938 it established a long-range forecast of research that would be necessary to any coherent war program. Among the eight areas of development within the radio field that were identified were radar, radio homing devices for planes, IFF equipment, and HF/DF. To a large extent, the research operations of the Radio Division in the first years of the war were dictated by this program. The NRL had recognized the importance of HF/DF, and it gave early and sustained support to its development.[30]

By 1939 the NRL was composed of administrative staff (mostly naval), seven laboratory divisions (all run by civilian heads and with civilian personnel), and two schools—the Radio Materiel School and the Interior Communication School, for the training of naval personnel. This unusual combination of naval and civilian control worked effectively and with a minimum of friction because "all concerned have realized that the laboratory is a service organization into which must flow renewed operating experience from the service itself, while scientific and technical continuity is assured by permanent civilian administration [of the research projects]."[31]

Acting Secretary of the Navy Charles Edison (who left office in 1940 to become the governor of New Jersey) continued his father's support for naval research. On 8 December 1939 he signed General Order No. 124, establishing the NRL as an independent entity directly under the secretary of the navy. The order made provision to continue the arrangement of joint naval-civilian control. Edison also appointed a senior rear admiral, Harold G. Bowen, to head the NRL, with the specific intention of increasing the importance and scope of naval research.[32]

Reflecting awareness of the war that had just started in Europe and the expansion of the U.S. Navy this was likely to trigger, Edison wanted "central organization" in the navy's research activities. On the same day that he signed General Order No. 124, Edison further rationalized the channels of communication with regard to scientific developments by signing General Order No. 130. The purpose of this order was to "emphasize research in the navy . . . [and] to effect a higher degree of coordination than exists at the present time."[33]

In order to ensure that the NRL would undertake work that was of direct benefit to the navy's operational forces, Edison appointed Admiral Bowen chairman of the Navy Department Research Council. This organization's members were drawn from the Bureau of Aeronautics, the Bureau of Ordnance, the Bureau of Engineering, and the Bureau of Construction and Repair (the latter two soon to be consolidated as the Bureau of Ships). The head of the NRL was also designated the technical aide to the secretary of the navy, charged with keeping the secretary up to date on research projects.[34]

A central information pool was created by transferring the research and invention duties of the Technical Division, office of the chief of naval operations, to the office of the secretary of the navy, where those duties came under the control of the NRL director. In addition, each of the material bureaus was ordered to designate an officer to head a section devoted to science and technology. That officer would act as liaison with the NRL and with the Navy Department Research Council.[35]

By placing so much authority in the hands of the director of the NRL, Charles Edison had not only emphasized the importance of research, he had also established a highly centralized system for the control of naval research. Yet it was also a system that, through its close links with the material bureaus, could be kept informed of the practical needs of the forces afloat.

Like his father, Edison continued to draw on the talents of civilian scientists to staff the laboratory. But he not only strongly encouraged a close interaction between the navy and civilian scientists, he also included private industry as the third essential ingredient in a productive relationship. Point six of his general order regarding the NRL reads in part: "The primary duty of the Laboratory divisions . . . will be fundamental research designed for the benefit of Naval Science and National Defense and such collected benefits to private industry as may develop therefrom."[36]

By July 1941 the Naval Research Laboratory was already heavily engaged in meeting the fast-growing operational demands of the navy; it seemed sensible to recognize its primary practical responsibilities. By this time Frank Knox had replaced Acting Secretary Edison, and he now moved the NRL from the office of the secretary directly to the Bureau of Ships. Knox abolished Edison's NRL-directed Navy Research Council and instead gave the NRL's former advisory functions to a coordinator of research and development.

At first the job went to Prof. J. C. Hunsaker, and later to Rear Adm. Julius A. Furer. Knox also gave the coordinator a new board with which to meet the greatly expanded tasks now emerging. This board was placed within the office of the secretary of the navy, theoretically "in order to secure a more complete measure of cooperation and coordination in matters of research and development, and to provide an agency for consideration of such matters."[37]

The coordinator's office was to act as a clearing house for progress reports on naval research from the various naval laboratories, commercial laboratories, manufacturers, and academic institutions. Naval officers were also deputed by the coordinator to provide representation in all significant civilian scientific groups, most notably the National Defense Research Committee, in order to secure close cooperation in the work on common wartime problems.[38]

What Knox had done was to move the NRL and its Research Council out of the research advisory business and out of the secretary's office to the Bureau of Ships, where it was better placed to concentrate on the practical technological work required by the bureaus in wartime. He had replaced it in the secretary's office, with his own research and development board and coordinator.

Though sensible in principle, in practice Knox's effort to further centralize control of research was largely ineffective because the NRL still retained its traditional influence. As far as the NRL was concerned, in fact, the administrative reshuffle was virtually irrelevant. As the NRL's war history explains: "NRL's relations . . . with the Coordinator were relatively insignificant."[39] The NRL continued to play the major advisory role given to it by the senior Edison and reemphasized by his son. At the same time, the result of the NRL's concentration on solving practical technological problems meant that during the war it was fulfilling the function for which it was best suited.

A good example of this is Maxwell Goldstein's work on HF/DF. As the head of the Direction Finder section of the NRL, Goldstein was handed the job of figuring out why it was so difficult to obtain readable bearings from the otherwise satisfactory FH3 and DAQ shipboard HF/DF sets. Goldstein theorized that the ship itself and its fittings were the problem, and that the best solution would be to minimize interference by locating the HF/DF antenna at the top of the highest mast.

But in order to test his theory Goldstein needed a seagoing warship, and, as he himself noted, "in the middle of war, seagoing warships were hard to come by." At this point NRL clout paid off: a destroyer was made available (briefly), tests were performed, and Goldstein's theory was verified.[40]

Much of the NRL wartime work was concerned with the development of equipment and devices for testing new naval gear, most of which, like ITT's HF/DF sets, was produced by industrial corporations. This kept the NRL at the heart of naval electronic development, in direct personal contact with service demands on the one side and civilian supply on the other.

The testing equipment was used to apply ever-higher reliability and performance standards to naval apparatus, especially for use in battle. As new requirements arising from service experience poured in from all over the world, the NRL conducted tests that enabled the bureaus to establish ad interim specifications to keep up with changing conditions. The close contacts the laboratory maintained with BuShips and manufacturers ensured that valuable suggestions for improvement of both specifications and material were incorporated into the final production decisions.[41]

ITT engineers who worked with the NRL on the development of HF/DF for ships attest to the closest possible relationship between the two organizations. Civilian scientists in both institutions apparently worked together with mutual respect and understanding. Thomas Edison's original concept of an in-house navy research center certainly paid off. And his son's support for a strong civilian role at the NRL made possible an almost seamless fusion of corporate, naval, and scientific interests cooperating to produce the best possible shipboard HF/DF.

In May 1943 the NRL instituted a practice that helped to assure the effective use in the field of the new technologies and weapons. Special traveling groups of officers and civilians were organized into an Electronic Field Service to supervise installation and calibration (where necessary) of new devices in the fleet and to advise on use and maintenance. The headquarters of these groups was at the laboratory, and they returned there periodically to keep up to date with new developments.[42]

The NRL had a considerable advantage over academic and industrial research establishments: it received constant informed feedback from serving officers. Used properly, these data most efficiently matched the technologies that were produced to the wartime demands. Yet the NRL did not have the capacity to handle all the requisite research, and it certainly did not have any production capacity to speak of, though it turned out many pre-production or breadboard models of operational equipment.[43]

The NRL's most efficient roles were just those it adopted in the course of the war. It became a conduit for practical navy demands, it provided liaison with the civilian academic institutions and manufacturing companies that could meet those demands, it established specifications for contracted production work, it tested and adjusted equipment produced outside, and it passed on to the navy the operating and maintenance requirements of that equipment. The principal product of the NRL was bound reports. The laboratory exerted its influence on technological and scientific decisions mostly by means of the text, engineering drawings, charts, graphs, and photographs contained in these reports.[44]

Fortunately, academic institutions and commercial enterprises were able and willing to fill in the scientific and technological research gaps that the NRL could not handle. By means of subcontracts, the NRL farmed out many phases of its projects to universities and commercial laboratories. A constant interchange of ideas was maintained and was encouraged by the many scientific committees and groups that emerged before and during the war.

It was not just the NRL, to be sure, nor even just the navy that encouraged, organized, and facilitated fruitful cooperation among academic scientists, industrial scientists and technicians, and the military prior to and during World War II. With the encouragement of Vannevar Bush, the president of the Carnegie Institute, Roosevelt had begun his contribution to this mobilization as early as June 1940, with an executive order establishing the National Defense Research Committee. The NDRC was to be a nondepartmental agency headed by Bush, with the task of coordinating research for both the army and the navy.[45]

Many scientific problems from then on, especially those connected with radio and radiation research, were no longer handled independently by the various army and navy and civilian research laboratories. Instead they were coordinated by the new committee into a cooperative effort to which all contributed. The committee established projects for the benefit of the navy at various universities; it also funded laboratories and hired scientists, engineers, and technicians. As the civilians knew little of the specific problems of the navy, every effort was made to provide proper orientation. This was usually accomplished by the operating agencies of the NDRC at the subdivision and section level. In April 1941, for example, a new section, C-4, was established to deal exclusively with ASW.[46]

The subdivisions of the NDRC committee worked directly with the navy materiel bureaus, as did the NRL, the other service laboratories, and the industrial organizations. When the NDRC was expanded in June 1942, Bush moved up to head the much larger Office of Scientific Research and Development (OSRD), but this did not significantly change the mechanisms for scientific development. Roosevelt brought in President Conant of Harvard University to take charge of the NDRC, providing his committee with an almost unlimited expense account. The NDRC became, in effect, the advisory panel of the OSRD, still maintaining its direct connection to ongoing research projects.

A bureau assigned a project to the NDRC, the NRL, an industrial laboratory, or any combination of them. Coordination was provided by frequent reports supplemented by conferences, personal liaison, and interlaboratory visits. In the field of electronics the NDRC's chief activities were concentrated at the Radiation Laboratory at MIT, and the Radio Research Laboratory at Harvard.[47]

Some contribution was made to direction finding, as to almost all other wartime scientific developments, by NDRC-sponsored projects, though the major part of the work was done by ITT and the NRL. In its own administrative history of the wartime years, the NRL lists high-frequency direction finders, along with pulse-identification systems, Loran, fire-control radar, and countermeasures against enemy radar, as devices "none of which can be claimed as the exclusive development of either NRL, NDRC, or American industry, but for which all three made significant developments."[48]

Certainly there were well-organized and pervasive structures for the

rationalization of scientific development during World War II. But neither the NDRC nor the OSRD, nor even the office of the coordinator of
research and development, made all decisions about the production of
weapons and devices for the U.S. Navy. High-frequency direction finding was one of perhaps many exceptions, the result of the inspired sales-
manship of Colonel Behn of ITT and of the NRL's longtime commitment to work in that field.

In fact, centralized organization and control of research and development for the navy was stronger on paper than in reality. The coordinator of research and development acted through the Naval Research
and Development Board, which spent much of its time keeping track of
the various divisions of the NDRC and trying to keep their efforts in
line with the primary needs of the navy. Wartime NRL director Admiral Van Keuren noted that while the office of the coordinator received
reports from all the navy laboratories, it did not do much about them.
The Progress and the Planning Sections of the Naval Research and Development Board were not very successful, apparently, in their stated
objective of the "formulation of coordinated programs of research," and
there was often duplication of effort.[49]

Nor did BuShips exercise more than nominal control over the NRL
during the war period. When Secretary Knox placed the laboratory under
BuShips administratively he retained for himself, through his Research
Board, the supervision of research. This reorganization also did not affect the issue of funding, although to some extent BuShips influenced
the nature of the research conducted at the NRL, because the bureau
supplied by far the largest part of the project money that supported the
laboratory. Congress gave the laboratory an annual appropriation that
covered running expenses and salaries of civil-service and contract employees for projects not chargeable to the bureau. But the predominant
influence of BuShips can be seen in the NRL's operating budget. By 1944,
at the height of wartime expansion, the laboratory's direct appropriation was estimated at about $2.5 million, while BuShip's projects at the
NRL (including much lesser work being done for the Bureau of Ordnance and the Bureau of Aeronautics), totaled about $10 million.[50]

Indeed, from the time the war started, the material bureaus were so
well supplied with funds that they could afford to sponsor pretty nearly
any idea that showed promise. By the same token, the NRL had sufficient financial backing "to undertake any research the overworked sci-

entific staff had the time and space to handle."[51] This added even more to its independence, and in practice the NRL made its own policy and administrative decisions, and on research matters it dealt directly with interested bureaus, private laboratories, and manufacturers.

Typically, too, there was only a small group of senior navy officers who circulated among the various boards, bureaus, and labs. They were managers with scientific and technical backgrounds, and they came to know each other well, so that the most effective contacts were usually direct and personal.

Early in 1942, Admiral Van Keuren had become the chief of BuShips. He quickly established a close liaison with Rear Admiral Bowen, director of the NRL (1939–42), having served with him for many years in the ship design bureaus. Later in 1942 Admiral Van Keuren moved over to head the NRL, where he remained until after VJ-Day. In the words of one of the past directors, "We were pretty senior admirals down there (at NRL), and the secretary and 'Ships' were far too busy under the pressure of war to worry how we did things, and they left us pretty much alone."[52]

By 1944, when James Forrestal had taken over from Frank Knox as secretary of the navy, the Navy Department realized that the various supervisory boards and councils established to coordinate all efforts had not been very successful. As a result, Admiral Bowen was placed in command of a new Office of Patents and Inventions. In May 1945 this was merged with the Office of Research and Development to form the Office of Research and Inventions, and by Executive Order No. 9635 the NRL was transferred there from the Bureau of Ships.[53]

It was, therefore, a relatively autonomous NRL that had conducted extensive tests on Henri Busignies's DAQ automatic direction finder for ships. It was the NRL's favorable reports that assured the production and deployment of Busignies's device in U.S. Navy vessels.

The Battle of the Atlantic was a close-run competition until the United States entered the war and began building ships faster than the U-boats could sink them. But those inexorable mathematics were far from clear at the time. What was clear was the constant pressure of the spiraling development of technology. In this war of technologies it may have been the side that made best use of its civilian scientific resources that had the winning edge.

CHAPTER 7

The Expediters: Procurement and Supply

Before World War II, few observers could have predicted the vast expansion that took place in American war production. During the First World War the production record of American industry was unimpressive at best. In the 1930s U.S. industry had been characterized by an almost total concentration on consumer goods, with no thought of planning for quick conversion in case of war. Japanese propaganda, with some truth, depicted the effete consumerism of the luxury-loving Americans. More practically, Dr. Joseph Goebbels, Hitler's minister of propaganda, was confident that the American railroad system, not having been planned for war, would not be able to move the vast amount of equipment needed for a war effort.[1]

Given the evidence before them, these attitudes were not unjustified. Many Americans in the 1930s, military and civilian alike, failed to grasp the significance of modern technology or to anticipate its vital role in a modern war. The federal government displayed serious deficiencies in this area. This was demonstrated by its prewar attitude toward high-technology transfer out of the country.

Avery Richardson, who later worked on HF/DF at ITT, recorded some startling examples of the parochialism and scientific illiteracy of the government. He pointed out that back in the 1930s, the State Department had heartily encouraged the company he worked for to take foreign orders for their commercial airborne radio compass. Even when an order came in from Communist Russia, the attitude was that "no harm could come from technically stupid Russians who certainly could not catch up with us."[2]

Late in the 1930s an order was received from Japan for the same device, and again the State Department blessed the deal. But Richardson

161

and his colleagues were "not comfortable with the prospect" of sending vital technology to an increasingly menacing Japan. So they filled the order as advised but left out a vital part, which meant that a bearing would look all right but would not function accurately.[3] This action was unusually prescient.

In the 1930s, isolationism was dominant in the United States. There were to be neither natural friends nor natural enemies. Distasteful foreign squabbles were to be avoided, and the country would trade separately and independently with all. On the outbreak of war in 1941, this attitude changed abruptly to an official xenophobia in the name of national security.

In retrospect it is sometimes difficult to understand, and even to remember, that involvement in World War II came as a great shock to many Americans. There had been a strong movement against any such entanglement. The military was kept underfunded in the interwar years, definitely not encouraged to maintain an aggressive readiness. Britain and France had gone the same way, and in 1939 had been similarly unprepared. But even with the conflict raging in Europe and the German U-boat campaign in the Atlantic affecting all shipping setting off from American ports, the U.S. Navy was still surprisingly unprepared for war when it finally came.

In 1940, in fact, the navy was still largely occupied with domestic concerns. Its offices routinely and politely fielded queries on radio communications from private individuals all over the country. The following letter from Paul R. Rockett of the Bronx, New York, is fairly typical of many in the records of the Bureau of Ships. Addressing his problem to the Radio Division of the Navy Department, Rockett wrote: "Gentlemen: Would you kindly advise me as to the trouble on my Philco Radio, Model 11, year 1928. Signal fades out at 15–20 mph speed of car, but the set continues to function."[4]

Only seventeen months later, at Pearl Harbor the navy would feel the first full effects of war, and its perception of radio electronics would change dramatically. Meanwhile, well into 1941, and in spite of the British example, the top U.S. Navy leadership did not seriously consider acquiring effective antisubmarine technology. Just as the Royal Navy had done before 1939, the U.S. Navy still put excessive faith in sonar.

In spite of slowly escalating involvement against U-boats in the Atlantic, American naval forces had not really been tested in battle when war broke out. They had little experience and would have to learn from their own mistakes. Even the unpreparedness that led to the early 1942 slaughter of merchant seamen by German submariners off the East Coast only slowly had the effect of encouraging naval scientific and operations research.[5]

By 10 December 1942, after one year of conflict, the chief of the Bureau of Ships reported in Circular Letter No. 26: "The Radio, Radar and Underwater Sound requirements of the various military services have placed a heavy load upon the radio industry. Production at present is just able to keep up with the more urgent demands."[6]

By 1943 the electronics industry in the United States was swinging into high gear. And by spring 1944 there were more than 22,000 officers and 225,000 enlisted personnel engaged in communications in the navy, most of them involved to some degree with electronics.[7]

Clearly, the military-industrial partnership that eventually took shape in the United States (unlike in Germany or Japan), was not the result of a well-planned strategy. Indeed, Roosevelt has been criticized for his "chaotic industrial mobilization effort."[8] The lessons about communications needs that had to be learned in the midst of the war were major. Both the military and industry were unprepared for the huge increase in the quantity of communications equipment, and for the heightened performance demands.

This was the first war to rely extensively on modern electronic devices. The trend toward obsolescence of technology was exponential, so that essentially victory came to depend on a nation's ability to design equipment; engineer it for production; produce it in quantity; and install, operate, and maintain it, all during the course of the war itself.

Rather than belabor the lack of advance planning on the part of the U.S. Navy, it is instructive to examine what interim measures they were able to put into place until the full energy of the country could be harnessed to the war effort. Before equipment could be produced specifically to meet navy requirements, the navy tapped a variety of other sources of supply.

Much electronic equipment was obtained by transferring it internally from one coast to another, from one shore station to another, and from one ship to another, wherever the need was greatest. Such transfers became common in 1941, mostly from the West Coast to the East, as the state of emergency grew closer and as the demands of the Atlantic Patrol and convoy-escort duty put great strain on the Atlantic Fleet.

On 14 July 1941, for example, the commandant of the navy yard at Puget Sound sent a set of transmission lines and fittings for a model CXAM-1 radio (radar) to the supply officer of the New York navy yard. The equipment had been intended for installation on the USS *Idaho*, destined for duty in the Pacific, but was then urgently required for the *Memphis* in New York.[9] By November the chief of BuShips was writing the commander in chief, Atlantic Fleet, that with regard to radar installations, "The deliveries to Boston . . . were based upon the directives of the CNO to stress early deliveries to the East Coast in preference to the West Coast."[10]

Some of the wartime supply practices that later became common, and that sensibly skirted paperwork requirements, were already in effect at this time. At the bottom of a June 1941 direction-finder shipment order is a note for the Bureau of Supplies and Accounts, which handled procurement finances. The note "requested that the Navy Yard, Puget Sound be authorized by dispatch to effect shipment in advance of receipt of this shipment order."[11] Gradually, with the step-up in vital production, such last-minute transfers became less and less necessary. Eventually new production, when not required for specific ships, went directly to local equipment pools from which it could be distributed as needed, thus cutting down on costly and time-consuming movements.

In the early days of the war, the navy was also able to make use of significant voluntary contributions of electronic equipment from industry and from private individuals. The navy bureaus received an avalanche of offers, many of which ended up in the files of BuShips, since it was in overall charge of procurement. Some volunteers offered equipment gratis, and almost all the rest wanted no more than nominal reimbursement. The following letter from the Radio Communication Service, Oakland, California, conveys the urgency of the times. It is addressed to the Chief of the Bureau of Aeronautics, U.S. Navy, and dated 20 December 1941. "Dear Sir: Do you need help in solving your radio communication problems in the quickest manner? We have transmitters, receivers and accessories in stock for immediate delivery. The enclosed outline describes our facilities. Please investigate what we have to offer and advise us how we may be of assistance to you."[12]

Before production in the United States could keep up with demand, supplies from abroad (generally Britain) were converted to American use. Reverse lend-lease swung into operation quickly in reaction to the

German U-boat assault on American coastal traffic. Eventually Britain sent twenty-four ASW trawlers and ten corvettes to help fend off that attack.

Maybe as another result of the U-boat offensive and at least partly because of some British successes in spring 1942, shipborne HF/DF at last came under consideration. But before the excellence of Dr. Busignies's high-frequency direction finder was adequately proven, and before it could be produced in volume in the United States, the gap was temporarily bridged when the NRL acquired a number of British sets (first the FH3 and later the FH4) and adapted them to American use for antisubmarine work. Eventually American versions were filling all of the domestic HF/DF needs. ITT'S DAJ, DAQ, and others were developed for naval use, and its portable SCR-291 was adopted by the Army Signal Corps and the U.S. Army air forces.[13]

Also prompted by the Paukenschlag emergency of spring 1942, the navy began to issue public requests for support. On 3 August 1942 Ralph N. Skrainka (who added under his signature: "Ex-USNRF 1918, honorable discharge") wrote to the Navy Department, Bureau of Radio Communication (*sic*) that:

> While vacationing in Cape May, N.J. several weeks ago I noticed a short article by Associated Press that the Navy Eastern Command was asking for used radio equipment to fill the gap while awaiting further commercial production . . . I have my entire station for sale as I figure the navy or Merchant Marine could use it *now* and I could further help by investing the proceeds in U.S. War Bonds. Then when the war is won I can again obtain amateur equipment. I have the following equipment . . .

Mr. Skrainka ended his letter: "With all the submarine sinkings it seems a shame that every American Merchant Marine lifeboat can not be provided with a transmitter and receiver."[14]

Another stopgap measure used by the navy to fulfill supply requirements before new production could keep up with demand was to request excess apparatus from the army. Much electronic equipment was secured in this way. On 20 August 1942, the assistant chief of naval operations for maintenance sent a memorandum to the assistant chief of staff (readiness) requesting army type SCR-511 portable radio sets. The memorandum states:

> In the event that the subject equipment is found suitable for Naval use, provision will be made for the following modifications to the stan-

dard Army design to increase the reliability under the conditions existing in the navy:

(a) Enclosure of spare batteries in sealed cans.

(b) Provision of waterproof fabric bags for storage and transportation of spare coil-crystal units.[15]

Avery Richardson did a lot of the testing of the ITT HF/DFs designed for the navy. He was particularly aware of the stringent requirements for sturdiness and waterproofing of electronic equipment destined for use in the adverse conditions of the Atlantic. Adapting land apparatus to shipboard use, however, was only an interim measure that could not adequately replace specifically designed equipment.[16]

In the summer of 1940, following the fall of France and the consequent threat of British collapse, President Roosevelt, always attuned to special navy needs, had initiated some measures to improve the country's readiness for war. On 27 June Roosevelt established the National Defense Research Committee, whose work was to enlist some of the country's best scientific brains in the cause of national defense.

This and other civilian bureaucratic groups the government established to oversee scientific research and development during the wartime emergency had considerable influence on the production of electronic equipment. The bulk of American electronic and radar research was conducted by MIT's Radiation Laboratory, Harvard's Radio Research Laboratory, and Columbia's Airborne Instruments Laboratory. These labs were created chiefly through the efforts of Vannevar Bush and the NDRC. Their impact was substantial, though indirect, as they did not actually engage in production. The government had to rely on private industry for that.[17]

Still, because of its large size, extensive budget, and power to grant production contracts, the NDRC was very effective in securing production of those devices conceived in the academic laboratories it had set up. The NDRC and the Office of Scientific Research and Development together put more than 32,000 scientists to work on war projects. This explains much of the incredible progress made in electronics between 1940 and 1945.

In fact, the suggestion has been made that the NDRC, because it had a specific, narrow focus, was more effective in encouraging the wartime development of technology than was the U.S. Navy, for whom so much of that technology was designed. It certainly did not hurt, too, that the NDRC was one of President Roosevelt's pet projects and had his wholehearted support.[18]

The War Production Board (WPB) also played a part in serving as liaison between civilian manufacturers and the military, and it helped to rationalize the supply process. The WPB organized Area Production Urgency Committees, such as the one for New Brunswick–Paterson, New Jersey, which oversaw the work of the ITT facilities in the region. Later, when BuShips granted clearances, supply contracts, and letters of intent to ITT, these were reported to the navy representative on the local WPB committee "for information."[19]

In the meantime, while implementing a variety of stopgap measures, the navy had also been reorganizing its own procurement system. On 1 July 1940, reacting to the threatening international situation, the Bureau of Engineering and the Bureau of Construction and Repair had been consolidated into the single, more efficient and centralized Bureau of Ships. But the navy administrators failed to anticipate the rapid development of electronics that was to take place in the next few years, and what had been the Radio and Sound Division was downgraded to become the Radio and Sound Branch, Design Division, BuShips.

In practice this reorganization gave the Bureau of Ships overall control of procurement for electronic equipment, including radar, sonar, and HF/DF, but in theory the financial arrangements were still in the hands of the Bureau of Supplies and Accounts. Although BuShips represented only one element in the initial decision to acquire HF/DF for the navy (Comdr. LeRoy Blaylock, representing BuShips, was present at all of the navy meetings with ITT), the bureau held final responsibility for the general disposition of the device. This control extended to the supply of direction finders to other services (including civilian organizations) and even to Allied forces.[20]

In the records of the Bureau of Ships at the National Archives, there are innumerable boxes filled with disposition orders for direction finders as well as for all other electronic equipment. Among them is a characteristic letter from BuShips, dated 2 January 1942, addressed to the Federal Communications Commission in Washington:

> In reply to your letter of December 15, 1941, in which you express an interest in certain high frequency radio direction finder equipment to be supplied the navy by the International Telephone and Radio Company, and request authorization to permit you to purchase certain of the equipments in question on an exchange-of-funds basis, the Bureau advises that such action will be satisfactory.[21]

In the beginning, officers were handicapped by lack of purchase and pricing experience. Most of all, until wartime appropriations could be implemented, they also lacked adequate funding to meet the navy's mushrooming demand for electronic equipment. Even the civilian contractors who were going to make the equipment did not know exactly how they were going to handle the manufacture of new devices, nor could they be sure of production costs. ITT engineer Avery Richardson expressed the dilemma inherent in the essentially new business of high-frequency direction finding when he noted that it was not possible to draw up exact specifications for a system that was "state of the art."[22]

Beginning with its own reorganization in 1940, and with the creation of the Bureau of Ships, the navy tried to keep pace with the almost overwhelming electronic equipment needs of all its forces on land as well as at sea. This effort accelerated rapidly after entry into the war.[23] Influence on the vastly increased production of electronics was chiefly channeled through procurement policies, contract negotiations, and grants of military service deferment. Industrial liaison officers and resident inspectors acted as field agents to supervise civilian electronic production. The navy's attempts to impose effective order on production were unceasing. They were carried on in an ad hoc manner in response to problems as they arose, which led to constant flux and change. While this was unsettling for many manufacturers, it also gave them considerable freedom to negotiate whatever arrangements best suited their own particular situation.

When James Forrestal had taken the oath of office as under secretary of the navy in August 1940, the first task Secretary Knox assigned him was to attack the mass of red tape and confusion that was tying up the navy's contract procedures. The naval expansion program was just getting under way at that time, and the instrument known as the letter of intent was quickly adopted to deal with the substantial increase in the volume of business. A letter of intent was an authorization from the navy to a contractor to begin work on an agreed project, with the understanding that the exact details of the arrangement, and sometimes even the price, would be worked out later in a formal contract.[24]

Before the war, naval procurement had been confined to only a few companies who knew the business well and who had the time to work through the vague, complicated provisions of the contract forms. But as the scope of the war broadened, the navy began to deal with literally thousands of producers, many of them inexperienced in government work and all of them under the most pressing time constraints.[25] The letter of intent enabled work for the navy to go ahead while, as Avery Richardson put it, "leaving the paperwork for later."[26]

The navy's ability to control its own purchase of equipment was extended on 27 December 1941, when President Roosevelt granted the military bureaus and offices the authority to negotiate contracts. This added even more flexibility to the system and shortened the process. Officers from the Office of the Inspector of Naval Material also acted as agents for the navy in examining businesses that wished to obtain navy production contracts. They assessed probable production capacities, measuring these against the often overly optimistic claims of the manufacturers.

Sometimes, as in the case of the December 1941 reports on the Teleradio Engineering Corporation, besides providing liaison between industry and the navy, the inspectors could also help manufacturers to match up with subcontractors. Of Teleradio it was reported: "It appears that this concern is not well setup to function as a prime contractor on navy communications equipment but do [sic] have suitable facilities and personnel to do a suitable job as a subcontractor making coils and subassemblies of radio receivers and transmitters."[27]

A follow-up report listed a number of companies engaged in producing radio equipment under navy contracts that might profitably use Teleradio as a subcontractor. Setting up conferences between such potential business associates was all part of the naval inspector's contribution to expediting the production of vital electronic equipment.

In January 1942 the Office of Procurement and Material was established within the Navy Department with, among other duties, the authority to act for the civilian War Production Board. This gave the navy a liaison with the WPB, which had the power to impose production regulations directly on civilian manufacturers. Navy inspectors could only issue reports. On 13 February 1942 the WPB advised the electronics industry that within four months, it had to be converted totally to war production. Radio manufacturers were forbidden to make radios and phonographs for civilian use after 23 April, and electronic-tube manufacturers were instructed to drop the production of nonessential items and to concentrate on those that were vital to the war effort.[28]

The issue of expediting the production of electronic equipment had now reached a critical stage, and, fortunately, Rear Adm. Samuel M. Robinson, chief of the Office of Procurement and Material, was both sensitive to the problem and flexible in looking for solutions. Admiral Robinson concentrated on creating practical guidelines to ensure a steady flow of equipment from civilian suppliers to the navy.

He recognized, for example, that changes in the design of equipment after it was in production resulted in delays. He noted that "we as engineers are inclined to regard changes as insignificant," but that if such changes, "no matter how slight from an engineering viewpoint, involve the procurement of additional materials or the construction of new tools, normal production procedure is bound to be interrupted."[29] Consequently, Admiral Robinson directed that "no modifications shall be made in the design of any of the above [radio, radar, and underwater sound] equipments after production is started and delivery schedules established unless these modifications can be introduced without delay."[30]

In the spring of 1942 another refinement in the navy's purchase of electronic equipment was undertaken, when a section devoted to electronics procurement was established in the Radio Branch of the Bureau of Ships. That section was henceforth in charge of all procurement.[31]

Later that year the bureaucratic structure was further complicated by the addition of another section, this one for "Progress," under the Production and Procurement Branch, Bureau of Ships. As its name optimistically suggests, this section was intended to enhance the navy's electronic production program. The navy's dependence on electronics was growing fast, and the demand for equipment was increasingly insistent.

The organization of electronics activity had in fact become so unwieldy that the Radio and Sound Division, which had been downgraded to a branch in 1940, had to be reinstated with division status in October 1942. By then some coherent advanced planning was finally possible with regard to naval electronic needs. The Radio and Sound Division was given a design branch, which worked with the office of the CNO to prepare manufacturing specifications for electronic equipment according to naval requirements.[32]

This latest attempt to rationalize and standardize research, field services, and procurement, though it lasted almost two years—from 1942 to 1944—was only partially successful in directing and controlling the supply of naval electronic devices. The U.S. Navy continued to increase its reliance on electronics at a furious pace, partly at least because of the influx of recruits with few sea skills who needed all possible technical assistance.

This was particularly true in the campaign against the German U-boats, which became more and more technology oriented during the course of the war. The navy's ability to cope with this expansion was always a case of running to keep up.

In order to head off serious supply snarls, Rear Adm. S. C. Hooper was made general consultant for radio, radar, and underwater sound equipment. His job was really to act as troubleshooter for the navy's procurement of electronic equipment. In July 1942, while assigned to the office of the CNO, Hooper had already pointed to a problem that might have hopelessly clogged the delivery of such equipment—the paperwork problem.

Even before the United States was actually at war, the Bureau of Ships had made requests for the shipment of urgently needed material ahead of the receipt of official shipment orders. This was only one of many ways of avoiding normal bureaucratic delays, and it was probably the navy's willingness to cut such corners that allowed the flow of equipment to continue and to grow rapidly, in response to the emergency.

The Office of the Inspector of Naval Material, particularly with the encouragement and help of Admiral Hooper, was especially helpful in making recommendations to the Bureau of Ships. This simplified paperwork requirements and rationalized procedures. After visiting radio plants in Chicago and New York in July 1942, Admiral Hooper had noted:

> In order to expedite new development models and to get production of these started promptly, it is recommended that the Army-Navy Munitions Board arrange that suppliers of materials and parts be given blanket authority to supply prime contractors with 10 percent of the material and parts necessary for the entire order upon receipt of the letter of intent if less than 1000 sets are to be ordered, or 5 percent if more than 1,000 sets.[33]

In some ways inhibiting restrictions on production were greatly reduced during the war. This is illustrated by the almost universal wartime use of the letter of intent to stand in for contracts, yet with all their authority to initiate production.

Admiral Hooper identified testing apparatus as another bottleneck in the manufacturing process. In December 1942 he noted that "whether it is discrimination or loopholes in the priority setup, inconsistent ratings had been assigned to testing apparatus which were particularly irksome for the small producer."[34] The practice of giving testing apparatus lower priority ratings than the final equipment to be produced often meant that manufacturers could not get the materials for their low-priority test equipment, without which they could not produce the high-priority equipment! This was less significant for the major corporations, which had the influence to obtain materials regardless of official priorities. However, it seriously hindered small producers who had to work through the regular, slow channels for priority assistance.[35]

Admiral Hooper also pointed out that competition between manufacturers and the armed forces was another factor detrimental to efficient supply procedures. "Much of the identical equipment is used by the Armed Forces as an end product," he wrote, "and by a contractor in the manufacture of testing apparatus." In this case the material ordered for the armed forces was classified as "supplies" and therefore got a high-priority rating, though the supplies might merely be stored for later use. The manufacturer received a low-priority rating for his testing equipment, and though it might have been required at once, he would not receive it until after the higher-rated orders had been filled. Admiral Hooper suggested that these problems could be resolved by giving the test apparatus the same urgency rating as the equipment it was designed to test. This would also keep the highest-priority AAA field open only for emergencies.[36]

There was yet another pressing problem that arose to impede industry's smooth production of war materiel: the manpower problem. The navy had some success in obtaining military deferments for vital companies, among them ITT's Federal Telephone and Radio Laboratories.[37] But this was never a blanket deferment, and ITT engineer Eugene Lombardi was transferred to the Army Airways Communications System, to set up an SCR-291 net in Northern Ireland, England, Iceland, the Azores, and North Africa.[38]

Because of the secrecy that obscured the importance of much of the work in electronics, the navy sometimes even had trouble protecting its own civilian technicians and engineers. Maxwell Goldstein of NRL's Radio Direction Finder section had at first been granted a class 2A deferment because of his work, but this was due to expire on 5 December 1941. In

May, Rear Adm. H. G. Bowen, Director of the NRL, wrote a letter to Goldstein for presentation to his draft board, clarifying his status as an employee at the lab. Admiral Bowen noted that Goldstein had been "working on a secret project in a highly specialized and complicated field." He told him that "to lose your services at this time would seriously retard the work in progress in connection with the National Defense."[39]

This letter was followed up by one in November from Comdr. R. P. Briscoe, acting head of the NRL, addressed directly to Goldstein's local draft board No. 24 in Washington. Commander Briscoe referred to Goldstein as a "key man" in charge of a "special development" at the lab, and an "outstanding scientist in this field of endeavor." In fact, he concluded, his services could not be replaced "by an equally competent relief at this time, since many of the results of this project have been the development of Mr. Goldstein."[40] These letters apparently did the trick, since Max Goldstein remained at the NRL until 1948. Not all other scientists were so lucky, however, and as the NRL war history notes bitterly: "Many draft boards were acting as if the war was being fought with sticks and stones."[41]

Here, once again, Admiral Hooper had some practical suggestions to make. When it was realized that there was a great need for more radio inspectors, he informed the chief of BuShips that "a number of valuable radio engineers were now in Censorship, where engineers are not essential." Displaying a common patronizing attitude toward women as well as the pervasive racism of the times, Admiral Hooper also advised: "White women are satisfactory in some details."[42]

At the end of 1942 Secretary Knox made yet another effort to streamline the procurement process. On 13 December he directed that each of the material bureaus (Ships, Ordnance, and Aeronautics) handle their own contracts for research, development, and procurement of technical items. Procurement of standard items of a nontechnical nature would remain with the Bureau of Supplies and Accounts. The Bureau of Ships had already been operating in this way with regard to electronic devices, so the secretary's directive merely legitimated an ongoing procedure. That spring, for example, Maxwell Goldstein, directly under contract with BuShips, had been testing the British FH3 HF/DF and running comparison tests with ITT's DAQ.[43]

Avery Richardson's experiences also highlight the considerable gap between rule-bound theory and actual practice. Richardson, who was on International Telephone and Radio Laboratory's HF/DF team and who often worked with Dr. Goldstein on the testing of HF/DF, enjoyed the war for the practical, freewheeling atmosphere it stimulated—a feeling that anything could be accomplished in the cause of victory.

Richardson remembers taking Busignies's direction finders as they were produced in New York, loading them into his station wagon, and driving all night down the Delmarva [Delaware-Maryland-Virginia] Peninsula and over to the Norfolk navy base in Virginia to "get the things installed."

"This is the time when the bean counters subside," he says, "when contract officers become flexible, when red tape is put away, all in the interest of getting on with it. Then the miasma descends again at the end of the war, but in the meantime everybody gets together and gets things done."[44]

Richardson and Goldstein were both exceptionally good at getting things done. They must have made a formidable pair—two civilian engineers, one representing the firm that produced a promising new HF/DF device, the other trying to get it to work practically on U.S. Navy vessels. Avery Richardson was tall and slim, with an irreverent sense of humor and irrepressible energy. He was a New Yorker, educated at Brooklyn Boys' High School where his father taught. After two years at Syracuse University, Richardson realized that he did not want an arts degree and transferred to the Brooklyn Polytechnic Institute. He graduated in 1924 with a degree in electrical engineering. In the following years he worked for several companies, including General Electric, before joining ITT in late 1940.[45]

Maxwell Goldstein described himself as an "unobtrusive man, short, tidy, and self-controlled . . . a civilian and a Jew."[46] Goldstein graduated from Johns Hopkins University in 1930, like Richardson with a degree in electrical engineering. His college yearbook describes him as "rotund, unassuming, reserved," and "always ready to lend a hand where it is most needed."

His practical, hands-on style was readily apparent to his classmates: "If you have a radio which refuses to click, or a 'gas buggy' that just won't turn over, page our friend Max. He has a reputation of making any radio talk, and the fact that his own 'hack' which should have been pensioned years ago, is still keeping pace with the march of progress, is ample proof of his ability in this line."[47]

Goldstein received his doctorate in engineering from Johns Hopkins in 1933, and he went on to develop radio navigation aids for the Army Air Corps at the Wright Field aircraft radio laboratory. He then worked on navigation applications for the Civil Aeronautics Authority before joining the Naval Research Laboratory in 1939.

During the war, Goldstein was in charge of the NRL's Radio Direction Finder section. Early on he also became one of three members of the Navy Department's Direction Finder Board, created to plan and integrate all naval D/F. In addition, he was very active as naval representative on the National Defense Research Committee for Radio Direction Finders. In 1943 Goldstein received special commendation from the director of Naval Communications and from the director of the NRL, as well as praise from the NDRC for his contributions to that group.[48]

This brilliant man, who had innumerable reports, papers, and patents to his name, never lost his early love of tinkering. Like Richardson, he reveled in just getting out there and getting things done. Charged with finding out why theoretically sound shipboard HF/DF equipment was not yielding workable bearings, Goldstein surmised that interference from the ship's own superstructure was largely to blame. With difficulty he arranged for the brief loan of a destroyer to test his hypothesis, but

> it took so long to mount the gear properly . . . that almost no time was left for the test itself. . . . The gear was rushed aboard on the afternoon of the day preceding the tests, and the entire evening and night was spent in setting up and tieing down gear with bailing wire, etc. During the early hours of the morning, the group retired to await sunrise tests. A Marine Guard was posted with strict orders to prohibit any molestation of the equipment. Tommie, one of our own investigators was left with the Guard as a further precaution.
>
> Just before sunrise, I was rudely shaken from my bunk with the startling cry that our seagoing warship had been torpedoed! Only after staggering to the pitch black porthole and stammering, "What are we doing?" did it slowly dawn on me that I was being told that the experimental setup had been "torpedoed" by the failure of the bailing wire, and that what was required was some quick technical assistance on deck to get the equipment operating before the skipper could get angry and heave us overboard with our gear.[49]

Goldstein continued his account, noting that though the captain of the borrowed destroyer became increasingly jumpy and impatient the longer their experiments took, this seemed nothing compared with the moment a depth charge inadvertently rolled off the fantail in the middle of a test. "But the theory was verified!" Goldstein was able to get usable bearings from Busignies's DAQ HF/DF while on a vessel at sea, and his accomplishment clinched the navy's decision to produce and deploy those devices with all possible speed.[50]

For his part in making U.S. naval HF/DF a reality, Secretary of the Navy James V. Forrestal personally awarded Maxwell K. Goldstein the navy's highest civilian service award. The citation, dated 5 July 1946, reads in part: "For distinguished contributions to the Naval Service in developing high frequency direction finding as a vital weapon for combating the German submarine menace during the crucial Battle of the Atlantic. These contributions came about because of his persistent research and development which led to the Navy's first successful shipboard high frequency direction finder."[51]

In spite of such generous recognition after the war, during the war the government continued to struggle valiantly to suppress the wild West sort of free-for-all that was so productive. In October 1942 the former Army Navy Communication Production Expediting Agency, which had been under army control, was reorganized into the Army Navy Electronic Production Agency, or ANEPA, and placed under a civilian director.[52] In 1942 ANEPA had a staff of 373, mostly from the army. The huge expansion in electronics, however, affected every agency, and a year later ANEPA employed over 1,000 people, more than a third of whom were in the field providing liaison with the more than 5,000 manufacturers who were by then engaged in electronics war work.[53]

It was partly the practice of extensive subcontracting that drew such large numbers of companies into war work, and many of the smaller firms would never experience official scrutiny. Some of the largest firms also profited from the emergency when antitrust laws were suspended for cases involving essential war production. The War and Navy Departments and the WPB had insisted on this measure. Indeed, the greater part of all navy electronic equipment was provided by only twelve manufacturers, among them the Federal Radio and Telephone Company. As well as coming under the supervision of WPB regional committees and naval inspectors of material, from October 1942 onward these twelve largest producers were routinely examined by local ANEPA representatives.[54]

What Richardson remembers, though, is not crippling restrictions on industry, but instead "no liaison people. When we wanted to talk to the navy we went to Washington. When the navy wanted to talk to us they came to Broad Street." Even contracts were not a problem. "Most of the time there was no question of arguing about modification of the contract. We were much more interested, in wartime, in solving the problem, and what to do next, and getting it done."[55]

Perhaps it was this spirit, more than all the bureaucratic efforts, that accounted for the success of electronics production during the war. Clearly, Richardson and his colleagues at ITT contributed in large measure to that success. During the nearly five years between 1940 and the end of the war in 1945, ITT fulfilled over 160 development contracts from the army, the navy and the NDRC. Most of these, including especially HF/DF, were of a highly secret nature.[56]

Apart from expediting production of electronic equipment by civilian companies, ANEPA was also charged with reducing competition between the various military branches for the services of the same firms. This was even a problem within the navy itself among the several branches of the Radio and Sound Division. There was a confusion of aims between encouraging competition among suppliers, in order to improve quality and cut costs, and discouraging competition for their products among the various military branches. This was a problem characteristic of the American system, which remained remarkably free market–oriented even in the midst of all-out war.

War-imposed security regulations, when added to the interservice and intraservice competition for production facilities, often caused the most serious delays of all. Both the army and the navy found that in order to obtain information on a device that was being manufactured by a civilian concern for the other service, it was necessary to go through channels as high as the secretaries of war and navy to clear the information. Sometimes this caused delays of up to three or four months before one service could get the details of an invention developed for the other.[57]

It has already been noted how security restrictions inhibited smooth cooperation between the army, the navy, and ITT in the development of Dr. Busignies's high-frequency direction finder. But on the other hand Avery Richardson, who was particularly involved in the testing and installation of that device, remembers things differently. He still enjoys citing a striking example of "one case of no red tape." In 1941—admittedly before the United States was officially at war—when the navy was building up its HF/DF coastal network, it wanted to locate its northernmost receiving station in Maine. Besides ITT and the navy, four other agencies had to be consulted on this project. All the participants, with Richardson and Busignies representing ITT, met one morning at Mount Casco, on the Maine coast. Since each was empowered to act on behalf of his organization, they had selected a site by 3:00 P.M., after which they enjoyed a fine, late lobster lunch and went home. Work on the site was started the next morning.[58]

No philosophical solution was arrived at to resolve the tensions inherent in a system that defied regulation, in a nation dedicated to individualism and personal autonomy. Such balance as was achieved was always precarious and depended more on temporary adjustments made by practical people than on the implementation of an effective organization.

The establishment of agencies, committees, divisions, and branches, both within the military and without, was one means of trying to secure the smooth and sufficient supply of electronic material for the prosecution of the war. But there were also many other control mechanisms developed to deal with specific problems as they emerged.

On 23 July 1942 Rear Admiral Hooper, then vice chief of naval operations, addressed a memorandum to the chief of BuShips on the subject of his recent visits to radio plants in Chicago and New York. Admiral Hooper reported that "competition between the army and the navy for critical materials is becoming serious to the point that a single priority list . . . is immediately essential."[59] As the navy orders were only about 15 percent of the total radio-manufacturing production, and as the navy needed the equipment for its ships and aircraft that were actually in contact with the enemy (which the army was not, at that time), Admiral Hooper argued that the navy requests should be given priority over those of the army.[60]

Failing that overall priority, Admiral Hooper suggested that instead of competition between the two services, there should be only one expediter organization, which should be governed by a highly specific priority list. In that case "it should be immaterial whether the senior expediter is Army or Navy." In the meantime, Admiral Hooper also urged the intensification of steps to increase production of those critical electronic materials, so that a priority list would no longer be necessary.[61]

While this must have sounded like a pipe dream in the summer of 1942, that point had actually been reached by late 1944, when technical equipment was being produced faster than men could be trained to use it.

In fact there were many different attempts, including at the highest levels, to impose order on the supply of electronics through the establishment of priority lists. In July 1943 Admiral King, then CNO, sent a directive to the vice chief of naval operations; the chiefs of the bureaus of Ships, Ordnance, and Aeronautics; and the coordinator of research and development, on the subject of the "Antisubmarine Materiel Program." Enclosed was a list of items broken down by priority classifications "as a guide to all agencies having responsibility for the development of antisubmarine material."[62] Shipboard HF/DFs in category A* (satisfactory state of production) were being delivered at a monthly rate of twenty, with a total order of five hundred and an allowance of one unit per vessel designated to be so equipped. These vessels were destroyers, destroyer escorts, Coast Guard patrol craft, and escort carriers.[63]

It was easier to establish an appropriate scale of priorities when the nature of the electronic devices to be produced was clearly understood. The NRL was heavily engaged in this clarification work for the Bureau of Ships. When the NRL improved equipment specifications and made them more exacting, the manufacturers were forced to test their own products more rigorously.

It was part of the responsibility of the resident inspectors of naval material to keep the manufacturers to whom they had been assigned informed of all these naval requirements, and to update them if they were modified. Avery Richardson noted that "the usual acceptance and test procedures are governed by the Navy Inspector who goes by the book and is rather suspicious that you are trying to put something over on him."

And, indeed, some contractors were, and did. But Richardson went on to point out that during the war, the situation improved because instead of inspectors armed only with book learning, "the Navy sent us three Chief Petty Officers of vast experience in our field," with whom he came to a good understanding.[64] While the inspectors were supposed to ensure that manufacturers met the requisite navy standards, there was also a diligent effort to see that such inspections did not halt the flow of supplies.

The practice of issuing regular reports on the state of production in the electronics industry was another effort to regulate and regularize supply. Such reports were issued by a number of different agencies and were distributed to other appropriate authorities. According to the Office of the Inspector of Naval Material in New York, ANEPA's "Weekly Production Progress Report for Electronic Equipments (ANEPA Form PB1) and the Monthly Production Forecast for Electronic Equipments (ANEPA Form PB2) are standard reporting forms for use by the producers of electronic equipments in submitting production facts and figures to the various interested departments."[65]

The reports were intended to provide another mechanism for the rational planning of electronic production. But such was the confusion inherent in the mass of competing military and civilian authorities that this aim was only imperfectly accomplished.

Manufacturers of electronic equipment had been required to furnish the Bureau of Ships with weekly progress reports on the ANEPA forms. They were also instructed to send a copy to the nearest ANEPA regional office. But when the Office of the Inspector of Naval Material in New York checked with companies in that district to be sure the reports were being sent to the Bureau of Ships, they were informed by several of the companies that they had received written notification from ANEPA that the weekly progress reports had been superseded. This was news to the Bureau of Ships, which believed it needed the reports. A handwritten note on BuShips's route slip attached to this correspondence makes clear the feelings involved (the emphasis is in the original note): "Apparently ANEPA issued instructions *direct* to mfgs without advising Inspectors in any way. This ties in with what Lt. Cmdr. Miles (Chicago) told me over the phone yesterday. Inspectors *have not* been advised as to the status of these reports."[66]

This exchange was taking place in March 1943, at the height of the U-boat onslaught against transatlantic shipping and well into the United States' second year of war. No one organization, military or civilian, had yet managed to obtain overall control of planning the production and procurement of electronic equipment for the U.S. Navy.

At the height of the war, in March 1944, Admiral Van Keuren, then head of the NRL, acknowledged: "It has sometimes been argued in the past that the interests of security and secrecy could be better served if the navy not only did all its research and development within its own service but some of the production work on especially secret apparatus as well."[67]

But the admiral rejected the idea of the navy taking over production of even the most sensitive electronic equipment on the grounds that this would inhibit what was a very close relationship with industry. This relationship was more important than were measures to safeguard secrecy, because it had enabled industry to "plunge into greatly expanded wartime production" with a thorough knowledge of the navy's "requirements, standards and special needs."[68]

It is quite likely that Admiral Van Keuren was aware that industry's standards of security were often tougher than those of some naval establishments. Avery Richardson was surprised at the ease with which he passed in and out of naval bases and receiving stations during the war, just by flashing a badge. This compared unfavorably with the tight security precautions at ITT's laboratories and plants.[69]

Yet, as Admiral King pointed out in his report to the secretary of the navy on 8 December 1945:

> It had often been predicted that in a national emergency the totalitarian countries would have a great advantage over the democracies because of their ability to regiment scientific facilities and manpower at will. The results achieved by Germany, Italy and Japan do not bear out this contention. Studies made since the close of the war indicate that in none of these countries was the scientific effort as effectively handled as in the U.S. The rapid, effective and original results obtained in bringing science into our effort are proof of the responsiveness of our form of government to emergencies, the technical competence of American scientists, and the productive genius of American industry.[70]

In the course of the war, the navy, government regulatory boards, and private industry were not always in agreement. Early in 1945, for example, ITT requested assistance from BuShips to help fund the acquisition of additional land in New Jersey, where the corporation planned to expand its facilities by building a whole new complex of offices and laboratories. BuShips enlisted the assistance of the War Production Board in determining whether or not ITT's work for the bureau justified the application for the $1,072,240 project. The WPB, in a telegram, replied in part as follows:

> Acquisition subject facilities clearly unnecessary for war effort. Present facilities located 67 Broad Street NYC considered by this office as sufficient to process orders on hand. It is believed that facilities listed

in Appendix A represent replacements and expansion which applicant would have undertaken if normal peacetime conditions had existed. Contractors opinions concerning percentage of production of proposed facilities which will be utilized for BuShips are vague. This office considers present proportion BuShips commitments of approximately 30 percent would not be materially changed. Present backlog at Broad Street laboratories $2,131,499 apportioned $712,562 for Navy, principally BuShips, $1,418,937, Army and NDRC. Approval of subject application is not recommended.[71]

ITT went ahead anyway and built its new facilities at Nutley, New Jersey. In 1946 they were opened by Colonel Behn with great ceremony, and for many years they were headed by Dr. Busignies.[72]

Starting soon after the end of war, navy spokesmen were generous in their praise of "the splendid support we had from the electronic industry and the many scientific groups in the country."[73] The wartime head of the Radio and Sound Branch of BuShips was especially appreciative, and he had certainly been in a position to know what was going on. "We have recently experienced in the course of a long and difficult war the wholehearted support of industry, scientists, engineers, and technicians of America, who gave unstintingly of their time and ability to build and equip the most powerful fleet the world has ever seen."[74]

But the electronic war, as waged by the United States, had been both "expensive and wasteful."[75] The American military thrust was to demand ever-better products without any break in the supply chain. To obtain this goal, development contracts for the same piece of equipment were often tendered to competing firms so that the Navy Department or the NDRC or OSRD could choose the best design. This sometimes led to duplication and even triplication of work. But it achieved the objective of obtaining an expanding supply of more and more advanced equipment.[76]

At war's end more than 1,200 people, military and civilian, were employed in the navy's Radio and Sound Division alone, which had expanded from the original 1940 total of 39. Between 1 November 1942 and 1 September 1945 the U.S. Navy had awarded contracts for communications or electronics materiel in the amount of $4.009 billion, and more than 550,000 workers in over 1,600 factories had been involved in filling those orders. Finally, on 20 August 1945, the WPB lifted the wartime ban on the manufacture of radios for civilian use. This was a fitting symbol of the end of the electronics war.[77]

By then the issue was no longer industry's ability to produce sufficient electronic equipment, but whether or not the navy could train its personnel fast enough to use the many sophisticated new devices.[78]

CHAPTER 8

The Operators: Installation and Training

In May 1940 R. E. Samuelson, an engineer from the firm of The Hallicrafters, Inc., in Chicago, sent a letter to the Navy Department, Bureau of Engineering, offering his company's services supplying marine radio equipment. Samuelson wrote: "I know that your usual navy requirements call for equipment that is easily tuned to any frequency in quite a large band."[1]

It was clear to a civilian engineer in 1940 that the navy's equipment must be as quick and simple as possible to operate. This requirement was doubly important in wartime, because the time constraints were so much more critical in battle and because of the need to train greatly increased numbers of men. Avery Richardson, also a civilian engineer, noted this same navy concern with ITT's HF/DF devices: "They [the navy] were always after us for improvements in operation and in speed of operation."[2]

In the early days of the war, many had hoped that superior technology would make up for the U.S. Navy's lack of experience. But the inescapable reality was that most electronic technology was virtually useless in the hands of inexperienced operators. In fact, given the success of the mobilization of American industry, the weak link in the chain from conception to utilization of most electronic equipment may have been the operator.

Looking back, a Navy Department analyst of naval wartime communication problems wrote: "I wish to make clear that our equipment was generally excellent. The amazing quantities of equipment produced, all of which met or exceeded the requirements of the specifications under which it was procured, stand as a record of achievement which is well known to all of you . . . [but] our personnel, because of necessity, were not always sufficiently trained."[3]

Between 1 November 1942 and 1 September 1945, the navy received electronic equipment worth more than $2.5 billion. Operators had to be trained to use this. But the problem was not only one of volume, although the huge expansion in normal communication demands was daunting enough. Innumerable varieties of highly skilled specialists were required, which swamped the regular training schools.[4]

The navy addressed the task of properly utilizing the vast output of electronic material supplied by wartime industry in two ways: by pressing the scientists to simplify their devices, and by increasing and improving facilities for training technicians. In the case of HF/DF, the efforts to develop simpler and more accurate equipment continued throughout the war.

A February 1944 NRL report addressed this problem. The objective of the report was "to investigate or to develop means for providing corrected H/F Naval shipboard direction finder bearings for frequencies other than those actually calibrated." These bearings had to be presented "in a manner that requires the least possible effort and skill on the part of the direction finder operator."[5] In spite of such intentions, high-frequency direction finding remained a complex art requiring the maximum in technical training and experience to prove useful operationally.

Three major schools for antisubmarine warfare were established primarily to combat U-boats in the Atlantic: the Naval Local Defense Force School in Boston, which had British-style mechanical teachers and concentrated on attack theory and procedures; the Fleet Sonar School in Key West, which taught the use of hydrophones and sonar; and the Submarine Chaser Training Center in Miami, where junior officers learned how to handle small ASW craft.[6] In addition, naval radio communications schools proliferated dramatically throughout the country.[7]

And still, in August 1942 Adm. Frederick J. Horne, vice chief of naval operations, expressed concern as to "whether present navy personnel is capable of handling modern radio equipment; that is, whether the equipment is becoming too complicated for proper operation and servicing."[8] This remark may have been inspired by the receipt of an 11 August memorandum that Lieutenant Commander Mallison of the Royal Navy dictated at the Boston Navy Yard. The "Memorandum on the Employment of High Frequency Direction Finding Equipment in Ships Employed on Anti-Submarine Operations" was sent to Commander Detzer in the Department of Naval Communications for circulation to "officers who might be interested, in the Naval Operations and Cominch Departments."9

Lieutenant Commander Mallison devoted a substantial portion of his memorandum to making recommendations about the training of operators in the effective use of HF/DF. He noted that to be of any value, direction finders must be continuously manned when at sea. This required an allocation of three operators per device, each of whom should have a good knowledge of German U-boat transmission procedures and should understand the meanings of the various signals received. That is, they should be conversant with the priority prefixes and should understand the significance of the length of the messages. It made a considerable technical difference, for example, if the signal received was a very brief first-sighting one or a longer follow-up informational one, which might indicate only the initial shadower in the vicinity. A second or subsequent short sighting report, especially one from a different direction, would indicate that more U-boats had come up.

Operators also had to be skilled in the difficult technique of taking D/F bearings. Emphasizing the unique role of the operators, Lieutenant Commander Mallison continued: "Being on his own when a bearing is intercepted, an operator must have sufficient judgement to report on the reliability of the bearing as well as identifying the type of signal. No one else in the ship can do this and the captain of the ship or the escort commander has to decide on the action to be taken on the verdict given by the operator."10

For these reasons Commander Mallison noted that Royal Navy operators, in addition to their regular radio training, received a two-week course of instruction at a radio interception station and spent another week at a school learning about the apparatus and the technique of direction finding. After joining a ship, they received as much practice as possible in taking bearings. Whenever practical it was arranged to have local HF transmissions made, preferably within sight of the ship, so that the operator could assess the accuracy of his own bearings and gain confidence in his ability and in the equipment, which, "with a false start may easily give bad results."11

Although no course could be given in this, it was also crucial for each HF/DF operator to become familiar with the particular sea-keeping peculiarities of his own ship. In order to obtain consistent bearings, the ship had to be held as steady on its course as possible, and the operator had to be aware of its behavior in different types of seas. Only time aboard could give that kind of knowledge.[12]

Mallison continued his description of prudent D/F practices, which the Royal Navy had learned by experience. In order to assist both operators and commanding officers, it had been found necessary to train a certain number of junior officers in the principles and operation of direction finding. They would then be able to continue the training and instruction of ratings while at sea. More important, the junior officers would be able to advise escort commanders on the particular value of each D/F bearing, something that required both skill and experience.[13]

It had been found that the best allocation of these junior officers was one to every four ships fitted with HF/DF. In this way they could coordinate the HF/DF organization in an escort group, particularly with regard to setting frequencies for watch keeping. These officers would also be responsible for the maintenance of the apparatus in the group, for checking calibration when required, and for preparing reports subsequent to an operation.[14]

In the Royal Navy operators were not even accepted into the HF/DF program until they had served a year at sea. HF/DF officers were given the same training as the operators, except that they spent additional time at the interception station and at the signal school. They also needed to demonstrate a "good sense of initiative, as well as good technical knowledge." Commander Mallison made it plain that the business of intercepting German U-boat high-frequency transmissions, determining the correct bearing, and interpreting the significance of the signal required a high degree of training of very able men.[15]

While it emphasized the training of operators, the memorandum also described the need for skilled D/F engineers to be available at every base where any number of D/F-equipped ships were stationed. These engineers would be in charge of all maintenance and spare parts, as well as undertaking calibration of the equipment. Calibration of HF/DF gear took two full days, and it could only be accomplished at certain selected bases.

It was necessary to choose a suitable stretch of water, free from traffic, where the ship could anchor at midpoint with a clear radius around it of at least a mile. Then a small vessel with a transmitter would sail around the ship, transmitting on the desired frequency. The HF/DF bearings would be checked against the transmissions, and all deviations were recorded. This procedure would be followed for up to thirty frequencies. They tried to carry it out on completion of the initial fitting of the equipment, and then at regular intervals of not more than three months.[16]

The U.S. Navy developed the practice of undertaking calibration of at least some of the HF/DF-equipped vessels at the outset of each task group's patrol. A typical record of this is the first entry in the daily operational narrative of the escort carrier *Guadalcanal*'s action report for May to June 1944. The entry reads: "13 May, 1205. Underway from Naval Operating Base, Norfolk, Virginia, to calibrate DAQ equipment in York Spit Channel."[17] Most CVE task-group action reports open with a similar entry.

In the conclusion to his memorandum, Mallison reverted again to the delicacy of the HF/DF operation. He pointed out that the success of the results obtained when a D/F bearing was taken at sea depended on four factors: "The care with which the ship has been fitted in accordance with the specified requirements; the results obtained on calibration; the degree to which the operators have been trained; and the state in which the apparatus has been maintained."[18]

There is no direct evidence that Lieutenant Commander Mallison's memorandum had any specific effect. But since June Commander Heineman had also been writing reports touting shipborne HF/DF. That was after his escort group, A3, had been able to drive off a wolf pack threatening convoy ONS 102 by using HF/DF bearings provided by the Canadian destroyer *Restigouche*.[19] Cumulatively, the information the U.S. Navy was collecting at that time suggested that HF/DF could be used effectively on ships, that the British knew how to do this, and that their success depended in part on taking special measures with regard to training in its use.[20]

Indeed, that same June when Commander Heineman became an advocate of HF/DF, an NRL report had recommended that "serious consideration be given to the extensive use of Naval shipboard high frequency direction finding on ground waves in view of the encouraging successes . . . *practically obtained* by the British."[21]

By fall 1942, the special measures Commander Mallison had advocated in August were indeed being taken, and training courses were put into place. These mirrored Mallison's memorandum very closely. Regarding the use of shipborne HF/DF, it seems that the U.S. Navy was willing to learn from the Royal Navy's experience.

On 27 October 1942 the commander of Task Force 24, based in Argentia, Newfoundland, addressed a memorandum to Vice Admiral Ingersoll, commander in chief, U.S. Atlantic Fleet. After describing the recent British successes with shipborne D/F, which made it an apparatus of comparable value to radar, he explained that HF/DF

> has repeatedly provided timely information to escort commanders which has been of inestimable value in detecting and attacking or driving off enemy submarines prior to gaining favorable position for attack.
>
> The ultimate effectiveness and efficiency of this new installation can only be attained when manned and operated by specially trained personnel. With this in mind and recognizing that the present allowance of radiomen in escort vessels is only sufficient for present needs, the following recommendations are submitted for consideration to the end that a policy may be formulated for manning the new HF/DF equipment both in Task Force TWENTY-FOUR and in the Atlantic Fleet as a whole.[22]

The first recommendation was that a school for training HF/DF operators should be established in the United States. Commander Mallison's lesson had been well learned:

> HF/DF cannot be manned, as can radar, with any degree of efficiency or satisfaction by men other than trained radiomen; reason—the HF/DF operators must know the radio Morse code, be able to recognize enemy procedure and peculiarities characteristic of enemy radio operators and tone of equipment, all to the end that he will not constantly furnish false or unreliable information based on signals that might be coming from friend instead of foe.[23]

This recommendation was followed up on 11 November, when Admiral Ingersoll sent a memorandum to Rear Adm. O. C. Badger, Commander Destroyers, Atlantic Fleet. Admiral Badger was notified that current planning was for the installation of high-frequency direction finding equipment in all destroyers as well as in other escort vessels. At that time the USCGC *Campbell* had a model DAR (converted British HF/DF) already installed and work was going forward fitting the USCGC *Spencer* with the same model.

It had been less than eight months since first approval for the testing of ITT's model DAQ HF/DF, and none were yet available for instal-

USCGC *Campbell,* date unknown. Note HF/DF antenna.
U.S. Coast Guard

lation. But they were in the works, and Admiral Ingersoll was aware that "the effective use of this new device requires due consideration as to personnel to operate it. . . . The Commander in Chief agrees," he stated, presumably referring to Admiral King, not to President Roosevelt.[24]

Admiral King's concurrence was demonstrated on 18 November, with a letter from the vice chief of naval operations directing the establishment of a permanent school, the Naval Training School (Direction Finders) at the Receiving Station, Mount Casco, Maine. There was pressure to get this school going as fast as possible, because delivery was scheduled to start by the following July (1943) on five hundred sets of American-made HF/DF equipment.[25]

In the meantime, a temporary school was established at the Navy Radio Receiving Station, Cheltenham, Maryland, which was already conducting classes for shore-based HF/DF operators. But chief radio engineer Lewis, the head instructor of that school, felt obliged to point out that "he did not consider that anyone now at the school was qualified to teach a shipboard material course."[26]

The problem of finding suitable instructors had not yet been addressed. Fortunately, it was learned from Commander Detzer at Naval Commu-

USCGC *Spencer,* 17 September 1943.
Note HF/DF antenna.
U.S. Coast Guard

nications that there were two persons in the navy, Ensign Kane and Chief Radio Mechanic Coyle, who had gone through the British HF/DF course. Ensign Kane was in South America at the time setting up HF/DF shore stations, "but it appears that his services could be spared." CRM Coyle was at a D/F station on the West Coast.[27]

It was recommended that Kane and Coyle be ordered to duty immediately to start the HF/DF school. And it was further suggested that three officers with suitable radio background and around eight enlisted men should be sent to England at once to take the HF/DF course there, with a view to coming back to head up the eventual American school in Maine.[28]

By the end of February 1943, two sessions of shipborne HF/DF training had been duly completed at Cheltenham. Robert J. Seroskie was one of those who went through the course. Bob Seroskie had earned the rank of radioman third class early in 1943, upon completion of the basic radio operator's course at Newport, Rhode Island. He was then sent straight to HF/DF training at Cheltenham, where

we were given the ABCs in the operation of the equipment which I felt were not unduly difficult to assimilate and did not require an inordinate amount of time. Beyond that we spent a larger part of our time in classes comprised of navy officers and some civilians where the subject matter was technical and, I might add, usually beyond the limited technical background of enlisted personnel, such as I, who had just come thru radio operators school.[29]

The temporary staff at Cheltenham prepared a curriculum based on their experiences (but perhaps without soliciting input from the students) for use at the permanent school that was ready to open at Mount Casco. The "Syllabus of the Course and Time Schedule, Shipboard Direction Finding Course, U.S. Navy" reflected the instruction that had been received from the Royal Navy, with the addition of instruction in the new American devices of Henri Busignies.[30]

The curriculum index indicates the heavy emphasis on theory, to which Seroskie objected:

> Reference Text Books [including Keen's still-classic work
>   on D/F]
> Fundamentals: Basic Requirements
> Wave Propagation
> DF Theory: Intermediate Frequency
> DF Theory: High Frequency
> Cathode Ray DF
> New HF/DF Equipment
> Installation HF/DF
> Safety Precautions

The procedures taught were German transmission procedure, logs and watch standing, Italian sub traffic, duties of HF/DF officer, and watches.[31]

The course was designed to last five weeks; the experience gained at Cheltenham indicated this was the minimum time required to qualify the average man assigned to the school. It was understood that men would already be qualified to receive eighteen words per minute, continental Morse code, upon reporting for the course. The practical part of the instruction involved a familiarization with the current HF/DF models, taking bearings on actual signals, simulated (dummy) transmissions, actual calibration of the device, use of testing equipment, location of trouble, and standing of search-receiver watches.

Seroskie does not mention this, but the official account indicates that a majority of the students trained at Cheltenham were Coast Guard

Radioman and HF/DF Operator Robert J. Seroskie, 1943, age 22. *Courtesy of Robert J. Seroskie*

personnel, because Coast Guard vessels were the first in the United States to have HF/DF equipment installed. Since the summer of 1942, the British and the Canadians had been doing most of the escort work in the North Atlantic. While still sharing operational control of the area, the U.S. Navy's actual contribution to North Atlantic escort was a mere 4 percent of the total effort.

The major U.S. commitment was to the Pacific and to convoying along the Atlantic coast, as well as to escorting troop and supply transports across the central Atlantic to the Mediterranean. The Royal Navy handled 50 percent of North Atlantic convoy-escort duties, and the Royal Canadian Navy covered the remaining 46 percent. Indeed, out of twelve Allied mid-ocean escort groups active at the end of 1942 and until the Canadians were withdrawn in February 1943, seven were British (with RCN, Polish navy, and Free French navy components). Four were Canadian (with RN components) and only one was American, made up of two U.S. Coast Guard cutters (Heineman's *Campbell* and *Spencer*), one U.S. destroyer, one RN corvette, and three RCN corvettes.[32]

Perhaps this explains Seroskie's next experience. It seems that there were not yet any navy vessels equipped with HF/DF when he and his colleagues finished the course.

Following our stint at Cheltenham in the spring of 1943 the navy decided that either it had more urgent need for our skills as radio operators or HUFFDUFF had not progressed to the point where it was ready for us. In any event three of us were assigned to a destroyer [USS *McLanahan*] as regular radio operators. We pulled convoy duty in the Atlantic and Mediterranean as well as participated in the invasion of

Sicily. This we did until early 1944 at which time we were given a two-week refresher course in the operation of HUFFDUFF at Portland [Mount Casco], Maine.[33]

The first class at the permanent school at Mount Casco began on 1 March 1943, under the direct administration of the commandant, First Naval District. The assigned task of the school was to provide adequate instruction for the training of 1,600 HF/DF personnel for the navy by June 1944.[34]

In addition to formal training at the school in Maine, HF/DF operators also profited from instruction at the hands of the engineers who designed the equipment. Avery Richardson recalls the role of ITT in explaining the vagaries of direction finding:

> There was a lot of work went into educating operators on interpretation of bad bearings. The skywave can come down with an odd polarization. It should have a vertical polarization but it bounces off the ionosphere and can come down even horizontal. So we could teach operators what the best bearing was. At Great River we had demonstrations. We had teams down to see how best to use the equipment. A lot of Chief Petty Officers came to spend time, for familiarization. We did not run a school, but we had instructors out to get their hands on the equipment, etc., so that they could then go back to train other operators.[35]

Before HF/DF could be put into operation, it had to be installed. The actual fitting of HF/DF devices in the ships of the Atlantic Fleet took place in a variety of different ways and on a flexible timetable. It has been noted that Coast Guard cutters were the first American vessels to obtain the device in the fall of 1942, largely because of the influence of Commander Heineman, and that this was a modified British model.

In May 1943 the escort carrier USS *Bogue* was fitted with a British HF/DF set during a two-week overhaul in Liverpool. The *Bogue* received HF/DF somewhat earlier than most other American vessels. At that time Benjamin Brooks was the skipper of the old destroyer *Belknap*, which participated with the *Bogue* in hunter-killer operations in the North Atlantic. Brooks remembers that although his ship had sonar and radar by 1943, it did not receive its HF/DF until 1944. By then the *Belknap* was participating in hunter-killer operations between Norfolk and Casablanca, with the escort carriers *Core* and *Croatan*.[36] Earlier in the war, the *Belknap* had seen service as a seaplane tender out of Hvalfjordhur, Iceland. Now it was a "born-again destroyer."[37]

Plans for the distribution of American HF/DF equipment called for installing it in approximately half of all destroyer-type escorts and patrol frigates, with the fitting being done prior to the commissioning of these new vessels. This accounted for the bulk of HF/DF installation, which began in July 1943. In addition, pools of the equipment were made available to fleet commanders for installation as desired in destroyers, escort carriers, destroyer escorts, and patrol frigates already in commission. This was usually accomplished when such vessels returned to port for maintenance. By 14 September 1943, Vice Admiral Horne, vice chief of naval operations, indicated that these plans were going forward smoothly.[38]

When Bob Seroskie completed his HF/DF refresher course in the spring of 1944, he was assigned to a World War I destroyer (USS *Tarbell*), which was in the Boston Navy Yard for overhaul. At that point it was also having HF/DF installed. The equipment was located in a small tailor-made shack on the main deck, a little past midships. The shack, notes Seroskie, "was similar to, but a bit larger than the conventional portable john" and barely had room for the equipment and a chair. Except for illumination from the scope, there was total darkness inside the shack.[39]

Ben Brooks adds that usually, "the antenna was on a separate mast about thirty feet tall. It was a diamond-shaped, four-wire box design mounted on a fixed axis. Visualize a box standing on one corner with an upright pole running through the opposite corner."[40]

On 14 September 1943, Admiral Horne expressed the by-then widespread appreciation of the importance of HF/DF, noting that "the greatest emphasis and highest priority has been accorded to HF/DF for the Atlantic Fleet."[41] But still there were many complaints about the slowness of the supply operation. The Combined Staff Planners report on "Measures for Combatting the Submarine Menace" had noted as early as March that "the HF/DF is new and highly desirable equipment for a percentage of vessels used in escort. The production and fitment of this equipment is slow and should be expedited."[42]

Apparently the delays continued, though, because in November 1944 an action report from the escort carrier *Mission Bay,* with Task Group (TG) 22.1, suggested somewhat plaintively: "Install HF/DF equipment

on all DE's. Only three of CortDiv [Escort Division] 9 have the DAQ."
Penciled next to this is a notation: "Being done."[43]

Securing an adequate supply of up-to-date HF/DFs remained an issue
of concern. But by the summer of 1943, when graduates of the HF/DF
school were beginning to man the new equipment at sea, the challenge
lay not so much in the development of new and better devices, nor even
in training personnel to use them, as in putting the devices into action
effectively so they could play a part in the outcome of the war. Effective
action depended on integrating information obtained from HF/DF into
the whole complex array of factors that made up a sea battle against the
German submarines. This had already been done with shore-based direction finding.

By late 1942 the U.S. Navy's network of shore stations dotted around
the Atlantic rim reached from Jan Mayan, a small island between Iceland and Spitzbergen, to Bahia in Brazil. There were also other stations
located on midocean islands, such as Ascension in the southern Atlantic. This network of direction finders continuously searched the airwaves
for U-boat broadcasts, noted their bearings, and then fixed the position
of the submarines by making triangulations from several shore stations.

The widely separated Allied receiving stations gave far more accurate locations than the Germans could obtain from their own shore stations, which were grouped too close together for good results. Naturally, the generally poor German HF/DF results encouraged the sense
of impunity with which the U-boats transmitted.[44]

Once a U-boat operating against Atlantic shipping surfaced and sent,
for example, a high-frequency fuel-status report to headquarters at Lorient (or later Paris, and finally Berlin), the broadcast would be picked
up by one of the Allies' shore D/F stations. This would instantly alert all
other stations and give the sub's radio frequency. All stations immediately tuned to this frequency and adjusted their D/F receivers for maximum volume. The "line of direction" of arrival of the transmission then
appeared as an ellipse on the cathode-ray screen, which was marked off
in the 360 degrees of a circle. The
direction of the ellipse on the
screen marked the direction from
which the sub's broadcast came.[45]

This direction, or bearing, was
sent at once to a "net control station," which forwarded the infor-

FH3 HF/DF equipment being tested
on *Corry*, May 1942. In action, this
equipment would be enclosed in a
small shack.
Naval Research Laboratory

mation to the U.S. Navy's Atlantic strategic high-frequency D/F net-plotting center in Washington. The center plotted such reports on a chart and determined the most probable location of the enemy from the intersecting lines. Naval Operations included this information, along with well-disguised intelligence from British and American cryptanalysis, in its daily U-boat summary, which it sent out on the Fox schedule, which was a prearranged schedule of times and frequencies for transmissions from headquarters. Such information could be used to alert escorts if indicated, as well as to reroute convoys away from danger areas.[46]

Before the creation of the Tenth Fleet in May 1943, the Anti-Submarine Warfare Unit, the Operational Information Section, and the Convoy and Routing Section, each integral parts of the headquarters of the commander in chief, might all be involved in collecting information for the Fox schedule and subsequently in coordinating antisubmarine activities. But even with the central direction accomplished by the Tenth Fleet, the daily U-boat summary was not always sufficiently current to direct successful anti-U-boat operations.[47]

Neither Ultra nor shore-based D/F could always be counted on to completely prevent confrontations between convoys and the large wolf packs Dönitz was able to gather together by this stage of the war, which were accurately directed onto targets by German decryption successes. However, the addition of shipboard D/F fixes gave an immediacy of time and a proximity of location that made it possible for escorts to launch direct attacks on those submarines that had managed to approach the convoys in spite of all evasive movements.

At this pivotal point in the campaign, spring 1943, it was of vital importance for the Allies to hit the U-boats with everything at hand. Captain Brooks explains the advantage of this: "The important thing is . . . that Huff-Duff enables escorts to pick up the ground wave of HF transmissions within a range of 25 miles. This enabled them to visualize the threat in terms of real time, rather than waiting for Lowell Thomas, or whomever, to come on the air with the daily summary from the sky wave hours after the transmission."[48]

Without having to wait for information from headquarters, HF/DF-equipped task groups could order escorts or planes or both to run down the bearing immediately on its receipt, in order to catch the detected sub before it was in position to attack or chose to move off. But it was not enough just to install shipborne HF/DF devices. In order to initiate an effective response to HF/DF bearings, it was necessary to persuade commanders to use the information provided.

This had taken time. Back in 1942, for example, during the night of 11–12 May, convoy ONS 92 had been attacked in the North Atlantic by one of the first successful wolf-pack operations. The joint U.S.-Canadian Escort Group A3 failed to prevent substantial merchant-ship losses in spite of the presence of the British HF/DF-equipped rescue ship *Bury*. This was at least in part because the bearings supplied by *Bury* were never used.[49]

Gradually, however, with training and experience, instances of the successful use of shipborne HF/DF increased. In November 1942, three U-boats approaching convoy SC 107 were located by the HF/DF on the Canadian destroyer *Restigouche* and driven off. And by March of the next year, British Comdr. Donald MacIntyre of Escort Group B2 used the HF/DF on his frigate *Whimbrel* so effectively that he was able to completely frustrate attacks on convoy ON 170 by a five-boat wolf pack.[50]

Building on early British experience and growing successes, the U.S. Navy gradually developed its own HF/DF devices as well as the doctrine for their use. Two years after Commander MacIntyre's success, in March 1945, a special operation codenamed Teardrop was mounted by the U.S. Navy on the direct orders of commander in chief, Atlantic Fleet, Adm. Jonas Ingram. Teardrop was the carefully planned response to a perceived threat to New York and other East Coast cities from V-1, rocket-launching U-boats.

When Ultra revealed that seven Norwegian-based U-boats of group Seewolf were headed for American coastal waters, the signal was given to begin the operation. Two unusually large mid-ocean barrier forces were created to block the passage of the presumably lethally armed U-boats. Each barrier force consisted of two escort carriers accompanied by more than twenty destroyer escorts.[51]

Capt. John R. Ruhsenberger, the commander of TG 22.1, has left a very detailed description of his part in what proved to be the last American antisubmarine operation in the Atlantic. Consisting of the *Mission Bay*, with VC 95 (naval composite air group 95) embarked, and accompanied by Escort Division 9, TG 22.1 conducted antisubmarine and barrier patrols between 27 March and 27 April 1945. Working in conjunction with TG 22.5 (with the escort carrier *Croatan*), Ruhsenberger set up the first search line north of the Azores across the known path of the advancing U-boats. By this stage in the war HF/DF was a recognized, significant part of antisubmarine operations, and its use was carefully structured and well integrated with the whole medley of weapons, tactics, and devices involved in a successful U-boat hunt.[52]

Before TG 22.1 left Norfolk, a conference had been held with representatives of OP-20, the Naval Communications Division. Based on recommendations put forward at the conference, a comprehensive HF/DF plan was set up for the task groups and incorporated into Enclosure E, the operations plan for the mission. Annex A (the communications plan) to enclosure E has an appendix 3, which is the D/F Plan.[53]

The D/F plan gives specific codenames for use in this operation. HF/DF was to be referred to as Jailbait (illustrating the intended effect of HF/DF, no doubt), and MF/DF was Banjo. Three frequency Guard Plans, Able, Baker, and Charlie, were set out—one for the northern group of ships in the escort group, one for the southern group, and a contingency plan Charlie. The frequency that each ship was to "guard" in each of the three cases was referred to by column number.

These numbers refer to a code where standard frequencies that were best received, or that the U-boats were known to use, had been assigned a prearranged column designation. Each ship's HF/DF equipment was supposed to have been calibrated for those frequencies according to the best information available at the time of calibration, just before the departure of the task group.[54]

By spring 1945, however, the Germans had become increasingly wary of transmitting, and when they did so they were using frequencies on either side of the normal spectrum.[55] So appendix 3, in addition to assigning a specific frequency for each vessel to guard, also assigned so-called off-frequencies to be watched. At 15 to 20 minutes, 35 to 40 minutes, and 55 to 60 minutes after the hour (the so-called silent periods) the HF/DF operator in each vessel was to guard the basic frequency assigned to him. During the "working periods," 0 to 15, 20 to 35, and 40 to 55 minutes after the hour, the operators were to guard the designated off-frequencies. These were expressed as a number of kilocycles, or negative, above or below the regular guard frequency. The plan was designed to sweep above and below the basic frequencies and was to operate twenty-four hours a day. In addition, if necessary, any ship could be ordered to take over any HF/DF guard station on any designated basic frequency.[56]

Fig. 8.1. HF/DF Guard Plan of TG 22.1, 27 March–27 April 1945.
NHC/OA, Action Report, Box 107

<u>S E C R E T</u> HF/DF GUARD PLAN

(a) Norddeich basic frequencies and Norddeich off-frequencies will be guarded in all areas. All ships will guard NFF tip-off. MISSION BAY and CROATAN will guard column Four (4) while operating East of Longitude 45 degrees West and other ships, if possible, will stand-by on column Four (4) in addition to regular Norddeich guard.

NORTHERN FORCES:

(1) The following DAU guard will be maintained during silent periods (15-20, 35-40, and 55-60 minutes after the hour):

	HOWARD FESSENDEN MENGES	BLAKELY FARQUHAR MOSLEY	JONES PRIDE	HILL LOWE
MISSION BAY Guard Column 4, Tipped off frequency, and check marking frequencies.	Guard Column 36	Guard Column 37	Guard Column 38	Guard Column 39

(2) During working periods (00-15, 20-35, and 40-55 minutes after the hour) the following off-frequency guards will be maintained.

HF/DF GUARD PLAN ABLE

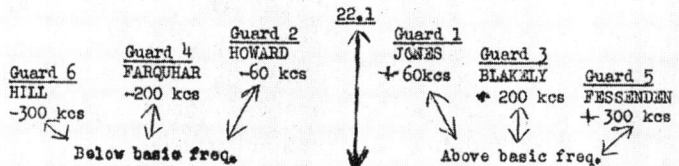

22.1

Guard 6   Guard 4   Guard 2   Guard 1   Guard 3   Guard 5
HILL   FARQUHAR   HOWARD   JONES   BLAKELY   FESSENDEN
-300 kcs   -200 kcs   -60 kcs   +60kcs   +200 kcs   +300 kcs

Below basic freq.   Above basic freq.

BASIC FREQUENCY (Col. 38 unless otherwise designated)

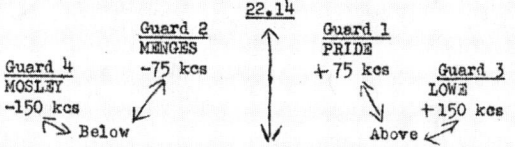

22.14

Guard 4   Guard 2   Guard 1   Guard 3
MOSLEY   MENGES   PRIDE   LOWE
-150 kcs   -75 kcs   +75 kcs   +150 kcs

Below   Above

BASIC FREQUENCY (Col. 39 unless otherwise designated).

-2-

All intercepted transmissions were to be classified as far as possible according to the character of signal—enemy, neutral or friendly; strength of signal, from 1 to 5; frequency band (low, medium or high); and origin of signal (ship or shore). These four signal characteristics were used to determine whether the intercept was to be classified as an emergency warning, a possible warning, or a signal of neutral or friendly character.[57]

For HF/DF operators an emergency warning was elicited by a signal of "enemy character, on HF ground wave from submarine, not from control." A signal warranting a "possible warning" classification was one of a "suspicious character, strength 3 or 4 signal, not above 600 kcs." As for "neutral or friendly" signals, if they persisted, they were located by obtaining a fix, if possible, so that the character of the transmission could be further classified.[58]

Apart from helping to locate friendly inbound planes when they were lost, the only signals HF/DF operators were concerned with were those of an emergency nature, which meant essentially any high-frequency transmission of German character received on a ground wave. This indicated the presence of a U-boat within something less than thirty miles. As operators became more experienced, they learned to identify such transmissions in a variety of ways. The following entry in an action report from the *Card* for 5 August 1943 notes a "good HF/DF bearing 168T, strength 3, on 9037 Kcs. Naval Enigma type. Submarine indicated by rough transmitter note."[59]

The elaborate HF/DF preparations for Operation Teardrop illustrates how important had become the location of U-boats by their radio transmissions. Even this late in the war, the high frequencies were guarded scrupulously for German signals.

In general, the communications officer on each American vessel equipped with HF/DF determined the watches to be kept by the three-man operator teams. The shifts were four hours on—alone in the cramped quarters of the HF/DF shack—and then eight off. Bob Seroskie explains that while on duty, the operator "maintained a log and monitored a series of prescribed frequencies, shifting from one to another as the watch proceeded."[60] At the end of each patrol, the logs were submitted "to OPNAV for subsequent analysis and for check of operators."[61]

The officer in tactical command was to be informed at once by TBS (Talk Between Ships; VHF voice radio with limited range, used for communicating within a task group) and in voice code of any intercepted traffic classified as emergency warning. Possible warning interceptions were to be reported visually during daylight for future investigation. Neutral or friendly traffic was not to be reported at all, unless there was some special reason for doing so.

The TBS circuit was guarded continuously and was the primary warning net for surface ships. It was to be used for transmitting urgent instructions or information too difficult to promulgate quickly by visual means. TBS was used "to transmit visual, radar, sound and HF/DF contacts . . . when speed and directness are primary considerations."[62]

In spite of all these elaborate and detailed preparations, the communications plan report for this part of Operation Teardrop concludes: "No definite ground waves were received by any of the units of TG 22.1 during the period in question."[63] Apparently some ship-based HF/DF operators had detected transmissions when the U-boats were still far to the north. But then, as it approached closer, group Seewolf observed strict radio silence, as did the barrier-force task groups. Finally shipboard HF/DF had nothing to listen to, but it was too late for the U-boats anyway. Eventually four of the group Seewolf were sunk with all hands, though the crews of the other three boats survived.[64]

Communications intelligence in the form of Ultra information had been extremely important in this last struggle against the U-boats, by making possible the systematic tracking of their progress across the Atlantic. In spite of the meticulously planned guard schedule, though, HF/DF had played a minor role. What shipboard HF/DF added to the war against the U-boats was another means of detection, but there were no panaceas. Shipboard direction finding did not replace fixes reported from shore stations. In fact, communications officers were specifically enjoined to "evaluate and advise Escort Commander of location of enemy vessels as determined by DF bearings obtained, often combining shore and ship DF bearings to determine the best possible fix."[65]

Nor was HF/DF, even shipboard and land-based HF/DF combined, sufficient on its own to locate U-boats. In practice, all the operations in the Atlantic needed to make use of every available means of location to pin down the elusive German submarines. Then they needed to use every available combination of weapons to secure a kill. By the time an HF/DF-dispatched escort or plane reached the U-boat's fix location, the sub had very likely traveled some distance and might have submerged. Radar would be used once an approach had been made to within range, and sonar could be operated if the sub was underwater. And still, throughout the war, one of the most effective means of locating U-boats remained the human eye. Even the position of a submerged boat could be estimated if a sharp-eyed sailor spotted the track of an incoming torpedo.

So, while HF/DF played a significant part in the Battle of the Atlantic, it was certainly no more a cure-all than was radar or sonar or even the most speedily available Ultra information. All these could do was to get the escorts to the right place. Offensive operations against U-boats also relied on weapons, vessels, tactics, courage, and experience.

But even defensive tactics were difficult enough. Diversion of convoys took a rare skill, or a great deal of good luck. As Ben Brooks saw it: "These warnings enabled convoys to make emergency turns which were sometimes successful, however hairy. Imagine directing a 70-ship merchant convoy to make two consecutive 45-degree turns at night while sailing under darkened ship conditions![66]

Ben Brooks, in fact, was left with an indelible impression that shipboard HF/DF was too demanding on the operator to be effective.

> By the time watch standers were aware of an incoming signal, there was virtually no time to take a bearing from one station, let alone make a fix from two. I do recall, however, that there was at least one instance in which the signal was so strong that the operator on another ship jumped from his shack and was able to point to the surfaced sub. I can't recall whether we got any definite signals on *Belknap*'s gear, but I do know that nothing was received that we could work with. If I can summarize, Huff Duff was a great idea, but was not sufficiently developed to alert tired watch standers to a contact given a split-second alarm span.[67]

The following entry in *Mission Bay*'s daily summary of operations is typical. On patrol between Norfolk and Dakar, Senegal, the *Mission Bay* report for 25 September 1944 notes "2127, HF/DF bearing on enemy transmission in column 10, strength 3 bearing 023 or 177; believed to be sky wave." The operator could not tell for sure either the sense of the contact, or whether it was a sky or ground wave.[68]

In spite of the difficulties, however, there is no doubt that the Royal Navy and the Royal Canadian Navy accomplished brilliant results against U-boats in the North Atlantic from late 1942 until Dönitz withdrew his packs at the end of May 1943. This was achieved in part with the assistance of their FH3 and FH4 high-frequency direction finders.

The story of the U.S. Navy's use of its own Huff Duff devices, beginning in mid-1943, is less well known. The navy successfully used HF/DF predominantly in the central and southern Atlantic and off the coast of Africa. Perhaps this has not been widely covered because it has been generally understood that the outcome of the U-boat war hinged on the North Atlantic run.

On 30 July Comdr. Rodger Winn, in charge of the Submarine Tracking Room of the Admiralty's OIC, assessed the current situation: "It is common knowledge both to ourselves and the enemy that the only vital issue in the U-boat war is whether or not we are able to bring England such supplies of food, oil and raw material and other necessaries, as will enable us, (a) to survive and (b) to mount a military offensive adequate to crush enemy land resistance."[69]

That being so, argued Winn, Dönitz was bound to return to the North Atlantic. Of course Winn was right, and when the U-boats did return that fall, the British and the Canadians resumed their relentless campaign using HF/DF and all the other weapons in their arsenal to successfully hold them at bay.

In the meantime, that summer of 1943 the focus had shifted to the central Atlantic, where the main burden of fighting the U-boats was carried by the U.S. Navy's hunter-killer groups.

CHAPTER 9

The Action: Huff Duff at War

I n January 1943 Roosevelt, Churchill, and their staffs met at Casablanca. Here, for the only time in the war, the campaign against the U-boats received top billing. It was obvious to most observers that future military plans, especially for the Allied invasion of France, could not proceed until the struggle going on in the Atlantic had been decisively resolved. By this time Dönitz could keep almost one hundred submarines, organized into large wolf packs, prowling along the convoy routes in mid-ocean, beyond the reach of Allied air. Something had to be done, and, finally, the Allied navies were given the resources and authority to step up their antisubmarine campaign.

Before new measures could be used effectively, however, changes had to be made in the irrational operational control structure. Direction of the Mid-Ocean Escort Force, which was responsible for herding convoys through the area of highest U-boat concentration, was split between the British and the Americans at the Change of Operational Control (CHOP) line. This line, at a longitude of approximately thirty-five degrees west, ran right through the heart of the most dangerous waters, which meant that overall control of the action often changed hands in the middle of wolf-pack attacks.

East of the CHOP line the British had centralized control of antisubmarine and convoy-escort responsibilities under the commander in chief, Western Approaches, with headquarters in Liverpool. West of the CHOP line, however, disorder reigned. Not only were there awkwardly overlapping American and Canadian sea and air authorities, but the U.S. responsibilities were divided up among competing areas. These included commander, Eastern Sea Frontier, who controlled U.S. escort and naval air operations from Florida to Maine; and commander, Task Force 24, at Argentia, Newfoundland, who controlled those activities between Maine and the CHOP line. To make matters worse, convoy movements were dictated directly from Washington, and there was virtually no coordination between the navy and the U.S. Army Air Force, which flew seaward antisubmarine air patrols.[1]

The large wolf packs operating in mid-ocean from 1942 onward caused the heaviest losses to merchant shipping in the whole war, and they finally alerted the Allies to an approaching critical climax. On the strength of the Casablanca mandate, an Atlantic Convoy Conference was convened in Washington on 1 March 1943, to reorganize the Allied effort against the U-boat in the Atlantic and to restructure convoy-escort responsibilities. At the conference it was agreed that most American escort groups would be withdrawn from the North Atlantic, and that convoys through that area would henceforth be protected by the Royal Navy and the Royal Canadian Navy, though for a time with some remaining American help.

The CHOP line was moved to forty-seven degrees west, giving the Royal Navy control to approximately the tip of the Grand Banks and thus responsibility for the ongoing wolf-pack battles. Canada negotiated full control of a Canadian Northwest Atlantic zone, stretching from the northern limit of the U.S. Eastern Sea Frontier up along the Canadian continental shelf. Adm. L. W. Murray in Halifax became the commander in chief with operational control of the Canadian Northwest Atlantic. The British increased the number of their escort groups in the mid-ocean escort force to ten B (British) Groups, and the Canadians were to return from support of the North African campaign to form four C (Canadian) Groups. The nominally American escort group, A3, was ultimately to be withdrawn. The U.S. Navy, under the direction of the newly formed Tenth Fleet, took over responsibility for the routes between North America, Gibraltar, and Morocco, and from the Caribbean to Britain, while the U.S. naval base at Argentia remained the headquarters for the U.S. Navy in the northwest Atlantic.[2]

At this time Admiral King, earlier the anti-convoy advocate of offensive action, was urging a basically defensive policy against the U-boats: convoy escort. But Adm. Sir Max Horton, the British commander in chief of the Western Approaches, favored more aggressive action. Sir Max, a "flamboyant" and "abrasive" man,[3] was a good match for Ernie King, who was reticent and abrasive. Sir Max had long been asking for the formation of support groups, which would be independent of convoys and thus free to track down and sink submarines. Finally, during the summer of 1943, with the ending of the convoys to Murmansk, the Admiralty could provide enough destroyers to form three support groups around the nucleus of the first three British escort carriers, *Biter, Dasher,* and *Archer.* At the same time, the Royal Canadian Navy provided a support group composed mostly of corvettes.[4]

In the summer of 1943 the Americans also organized offensive task forces—hunter-killer groups—around their own escort carriers. Each country retained operational control of its own forces operating within its own command areas. Now the tide of battle turned against Dönitz and his German submarines as these three groups—the American, the British, and the Canadian—were each equipped with shipboard HF/DF, centimetric radar, and new antisubmarine weapons. The Allied antisubmarine campaign was further strengthened by the following three developments: (1) the closing of the mid-ocean Air Gap by the provision of sufficient numbers of VLR aircraft; (2) the establishment of an improved command structure by the Washington Convoy Conference and the creation of the U.S. Tenth Fleet; (3) intelligence from cryptanalysis and land D/F.

The support and hunter-killer groups were not meant to search wide areas of the ocean in the hopes of finding U-boats. That had been tried by the Royal Navy early in the war and was fruitless. Instead, it was recognized that to be effective, the groups had to be furnished with intelligence from Ultra about the presence and approximate location of the enemy submarines. Operations were to be coordinated with and linked to convoy activity, as that was where the U-boats would be found. Ultra was the first instrument to send the carrier groups into the right areas. Then, when the U-boats made transmissions, shipboard HF/DF would come into play.

The Royal Navy's support groups soon achieved some notable successes against the formidable North Atlantic wolf packs, and HF/DF played a significant role in many engagements. Between 22 April and 6 May convoy ONS 5 had been dogged, harried, and attacked by a fluctuating number of U-boats from among the sixty or so in the vicinity. The convoy was escorted by Escort Group B7 under its highly skilled commander, Peter Gretton, and it was eventually also protected by elements of the 3d and then the 1st Support Groups. While radar was employed to great effect in the foggy weather that engulfed the convoy, HF/DF had also been in constant use in this running battle, giving warning of approaching U-boats. Twelve merchant ships were sunk by the end of the ordeal on 6 May, but the U-boats had lost six of their own number. They withdrew "shaken and demoralized."[5]

A few days later some thirty-six U-boats failed to press home attacks on HX 237, which was strongly supported by aircraft from the escort carrier HMS *Biter*. Three merchantmen were sunk, but so were three U-boats. The two destroyers and four corvettes escorting nearby convoy SC 129, though unable to prevent the early loss of two merchant ships from the convoy, managed to sink one U-boat and severely damage several others, keeping the rest of the strong pack at bay. Referring to this last action, Dönitz wrote in his war diary:

> No less than eleven of the boats in contact with the convoy were detected and driven off while it was still light. This is a very high percentage. It is obvious the enemy must have detected all the boats in contact with astonishing certainty.[6]

Of course Dönitz did not know that the Allies now had an effective shipborne HF/DF. Even while wondering how it was that as soon as a U-boat approached a convoy it was detected and driven off, the Germans never made the connection between their radio convoy-sighting reports and the sudden appearance of an escort.[7]

In the third week of May 1943, extensive use of HF/DF again helped protect a convoy from aggressive attack. HX 239, with forty-two ships, had sailed from New York on 13 May and was joined by Escort Group B3 on 19 May. Two days later the convoy was also joined by the 4th Support Group with the escort carrier HMS *Archer*. On 22 May the escorts began to pick up numerous HF/DF bearings indicating U-boats close by.

When a plane was sent to chase down an HF/DF bearing, it unexpectedly sighted what turned out to be convoy ON 184, about thirty miles to the north heading west. The U-boats of the Mosel group were in the unfortunate position of being caught between two convoys, both with carriers in attendance. HF/DF directed aircraft from the carriers down the lines of bearing directly to the U-boats, two of which were sunk, and several sustained damage. Located by HF/DF and held at bay by aircraft and escorts, the U-boats were prevented from attacking the convoys. Dönitz broke off the action against HX 239 on 23 May.[8]

Earlier that month the escort carrier USS *Bogue*, under the command of Capt. Giles E. Short, sailed from Belfast with the 6th Support Group in fulfillment of the Atlantic Convoy Conference agreement to continue for a short time to provide some assistance on the North Atlantic run. The *Bogue* was detailed to provide air cover for convoy ON 184 from the Iceland area to Argentia, and the commander noted in his action report that "the HF/DF equipment proved invaluable." This was probably the first time HF/DF had been used successfully by the U.S. Navy, and the device was a British one.[9]

A continuous watch was kept on the newly installed equipment from the time the *Bogue* left port. The commander in chief, Western Approaches, had assigned Sub-Lt. J. B. Elton, Royal Navy Volunteer Reserve, to assist on the trip by supervising the three American radiomen from the *Bogue* who manned the HF/DF. On that morning of 22 May, when ON 184 and HX 239 approached each other, the equipment was put to its first use—to home in a returning TBF-1 Avenger aircraft.[10] This turned out to be a subsidiary but valuable application of HF/DF. The use of the device to home in carrier planes became widespread, and it undoubtedly saved many lives and much equipment.

At 1051 (local time) that same morning an HF/DF bearing was obtained on a U-boat, but before the bearing could be transmitted to the escorts, a TBF on patrol had already attacked the sub. That afternoon, an HF/DF bearing was directly responsible for an attack on a U-boat resulting in its surrender. At 1727 a bearing was obtained on a fifty-nine-group Enigma message, on series 8, from the vicinity of convoy ON 184. This was believed to be a report from a U-boat notifying headquarters that it had sighted a convoy. Because it was so long, the message was undoubtedly the follow-up contact report with full convoy and escort information. Planes from the *Bogue* flew down the line of bearing and attacked the still-surfaced submarine, the U-569, which was forced to surrender and scuttle. German survivors were hauled from the water and taken on board one of the escorts.[11]

The action-report narrative continues, explaining that in all, five attacks on four U-boats were made by *Bogue* planes in the course of the afternoon. Apart from sinking the U-569, which was the first kill for a U.S. vessel in which HF/DF was involved, two of the other subs attacked were reported as "probable sinkings." In fact, however, they escaped.[12]

Captain Short's narrative concludes that "without doubt the latter transmission at 1727z [local time zone] was made by the U/B [which surrendered] and wrote its death warrant." He also points out that that same transmission had been picked up by shore D/F, notice of which had been sent to "ships escorting HX 239 and ON 184."[13] It is very apparent that the combination of land-based and shipboard HF/DF placed transmitting U-boats in an extremely vulnerable position; one that could be exploited immediately by escorts and carrier planes. Convoys HX 239 and ON 184 escaped unscathed.

A few weeks after this success the *Bogue* was shifted to the New York– Gibraltar route, to begin a more aggressive strategy directed by the Tenth Fleet. This followed the recommendations of the Combined (Allied) Staff Planners in a report of 1 March, and it reflected the offensive ideas that Adm. Max Horton's groups were already putting into practice. The report had stated that recent enemy successes should be met by, among other things, the "provision of striking forces of surface and aircraft, additional to those needed for normal convoy escort, to reinforce threatened convoys and act offensively against the U-boat concentrations."[14]

Forty-nine years later Ben Brooks, who had been on the *Belknap* operating with the *Bogue* against the Mosel group, described how those new tactics appeared to a young destroyer commander:

> When I had the ship [the USS *Belknap*] we'd come down from overhaul here in New York, and pick up the carrier at the end of the swept channel off the Virginia Capes, and we'd screen and escort to Bermuda. We'd top off with fuel and stores and everything in Bermuda and then go out and just mill around. If a convoy was coming this way and another convoy was coming that way we'd be out there. But we'd try to stay just over the horizon, though we'd be within calling distance of the convoy, and ready to close on anything that turned up.[15]

Brooks and many others like him were out there that summer of 1943 because the stiff, mostly British, opposition had persuaded Dönitz to withdraw his boats from the North Atlantic at the end of May. Dönitz moved the bulk of his remaining U-boats to the central Atlantic to prey on convoys steaming from the United States to the Mediterranean and tanker convoys heading from the Caribbean to Britain. He also stationed his U-tankers in this area, especially to refuel U-boats heading farther afield, to the south Atlantic and to the Indian Ocean. The *Bogue* was soon joined by three more escort carriers, the *Card*, the *Core*, and the *Santee*, accompanied by covering hunter-killer groups. In June, July, and August, these groups mounted a concerted effort against U-boats in the vicinity of the Azores.

During those months planes from the escort carriers sank sixteen

U-boats, eight of them tankers. The result was not only to protect shipping in transatlantic convoys, but also to disrupt U-boat operations by preventing refueling. The tankers were a particularly vulnerable target because of the great number of radio transmissions required to set up the refueling arrangements.[16]

On 31 May 1943 the *Bogue* and escorts *Clemson, Osmond Ingram, George E. Badger,* and *Greene,* comprising TG 21.12, had sailed from Newfoundland to take up their new station northwest of the Azores at latitude forty degrees north, longitude fifty degrees west. According to their orders the group was to "operate offensively" in support of "African convoys." Information was furnished to the commander of the task group, Captain Short, about the position and routing of certain convoys, GUS 7A (from Gibraltar to New York) and UGS 9 (from New York to Gibraltar) being the ones of most immediate concern.[17]

Captain Short realized that he had a twofold mission, both to patrol areas of "estimated possible concentrations of subs" and to guard areas in the vicinity of the convoys for which he was responsible, which might be threatened. It was also clear that TG 21.12 was sailing into a fluid situation, subject to continuous change. The theater of action was large, and it was certain that the reported U-boat concentrations were not going to cooperate by remaining in place for very long. Captain Short saw the challenge as one of ensuring that his task group was in the right place at the right time. In his own words, his mission was to "operate offensively."[18]

In order to be ready for action at any time, one of Captain Short's major preoccupations was to see that his escorts had enough fuel. This was an immediate and continuing concern, as the destroyers had a small fuel capacity and therefore a relatively short range. The objective was to keep the escorts topped up as far as possible, from convoys when this could be done, or from the *Bogue* when necessary. In order to ensure that the escorts were self-supporting, they had to have enough fuel either to reach port or to intercept a convoy if the *Bogue* should be sunk. Having made these preliminary judgments, Captain Short proceeded with his patrol. Early on 4 June aircraft from the *Bogue* attacked a U-boat (U-228), forcing it to submerge. Later that same day, an HF/DF bearing indicated the possible presence of another submarine to the south of the one that had been attacked in the morning. Aircraft were dispatched to the location, but this time they did not find anything. The "Schedule of Flight Operations" notes that at 2048, "launched TBF to check on

HF/DF bearing 353 degrees. Did not sight."[19] Two other U-boats were encountered by planes from the *Bogue* that day, though, and they had both been forced to submerge as well, but not before one of them (U-641) had shot down one of the aircraft. Because the number of HF/DF fixes indicated a continuing danger, it was decided to move south and to make contact with convoy UGS 9, in order to provide it with direct cover.[20]

On 5 June two aircraft of VC9 from the *Bogue* sank the U-217. The support group continued on its southerly course, making a wide search of the waters around the seventy-four-ship convoy UGS 9 and its escorting Task Force 69. The "considerable HF/DF activity" noted by Captain Short made it certain that the convoy was still being shadowed.[21]

The best way to counteract shadowing was to continue forcing the U-boats to stay submerged. This was supposed to prevent them from maintaining contact with the convoy; at least it would make it very difficult for them to get into a good position from which to launch an attack. The continuous presence of escort vessels, as well as sustained patrol by the carrier planes, was the most effective way to keep subs down.

Since they had first been located by HF/DF and attacked on 4 June, the U-boats in the vicinity of convoy UGS 9 had become very wary and seldom surfaced. When they did, they invariably sent out a radio report. Some of these transmissions, mandated by headquarters, included sighting reports of aircraft from the *Bogue*. But use of the radio brought the planes back, and sighting the planes forced the subs to submerge again. With the combination of sufficient escorts (four of his own and five in the convoy task force) and carrier planes, and alerted by HF/DF, Captain Short could maintain tight cover of a large enough area to prevent any dangerous movement of the submarines. Because of their slow underwater speed, shadowing U-boats could only move up to an attacking position on the surface, which they were understandably reluctant to do with planes buzzing all around. It was not until 8 June, therefore, that another action took place. At 1500 on that day, a TBF piloted by Lt. L. S. Balliet was sent out to investigate an HF/DF bearing at 242 degrees. At 1706 he found a surfaced sub (U-758) about ten miles from the convoy on a bearing of 220 degrees. The TBF attacked the sub with four depth charges. When the sub fought back with antiaircraft fire (its new 2cm quadruple gun as well as its two twin 20mm mountings), two TBFs and one F4F fighter were sent to assist Balliet.

One of the newly arrived Avengers attacked the sub at 1747, dropping three depth charges. The plane received several 20mm hits from the German guns, which wounded the radioman and badly damaged the engine. Nevertheless, the pilot succeeded in landing back aboard the *Bogue*. At 1828 the F4F Wildcat made four strafing attacks on the sub, and the remaining TBF attacked with four depth charges.[22]

The account of this action in the *Bogue*'s schedule of flight operations concludes by noting that the "submarine was at least badly damaged."[23] This estimate, as often happened, was overly optimistic. In fact, "heavy antiaircraft fire and smart, cagey tactics saved U-758," which was "merely shaken" in the action.[24]

HF/DF had picked up a characteristic U-boat B-bar message from the vicinity shortly before the U-758 was attacked by Lieutenant Balliet. Because this had occurred well within visual contact of the convoy, Captain Short noted in his report of the incident that "it is believed this sub made a sighting report on the convoy." Captain Short feared that the report had gone to U-boat headquarters, where it would have been retransmitted to alert other submarines in the area as to the exact location of the convoy.[25]

No further U-boat sightings took place that day, but immediately after the recovery of the four planes, considerable HF/DF activity was again noted. Together with the previous B-bar message, this activity indicated to Captain Short that the convoy might be in danger of a pack attack. No attack developed that night. But because of the HF/DF interceptions it seemed a "probability" that a concentration of U-boats was forming ahead of the convoy for an attack on the next night or the night after. In consequence, a continuous daytime sweep was maintained ahead and on the flanks of the convoy.

At 0748 on 9 June, an F4F was sent to investigate an HF/DF bearing at 355 degrees. The plane flew down the line of bearing for fifty miles without sighting anything and then joined the regular patrol. There were no further sightings that day, and on the next day, 10 June, Captain Short's merchant charges came within range of shore-based PB4Y patrol bombers. At that time the task of protecting convoy UGS 9 was handed over to the Liberators, and TG 21.12 moved out to hunt for U-boats on its own. Because there were no sightings of U-boats on 11 June or early on 12 June, it was obvious to Captain Short that the submarines were very alert to the danger from aircraft. They were either remaining submerged, or they were surfacing only long enough to transmit brief, required messages. Finally, at 1347 on 12 June, a U-tanker was sighted (U-118) and sunk by eight aircraft from the *Bogue,* about twenty miles from the carrier. It was not recorded if HF/DF bearings played any part in this action.[26]

On the evening of 14 June, at "about 1938 . . . an HF/DF bearing was received." A plane was dispatched about fifty miles to the northward along the line of bearing, but no submarine was sighted. By that time the *Bogue* had to have repairs made to its catapult. Some of the destroyers were also in need of repairs and the patrol was almost over anyway, so Captain Short received permission to proceed to Hampton Roads. He had successfully accomplished his mission.[27]

On 27 July 1943 at 1158 the escort carrier *Card* got under way from Norfolk, Virginia, with three destroyers, *Barry, Goff,* and *Borie,* of TG 21.14. Under the command of Capt. Arnold J. Isbell, the task group's assignment was to conduct air search operations against enemy submarine concentrations south of the Azores while en route to Casablanca, and then sweep back to Norfolk. In addition TG 21.14 was to provide air support for convoy UGS 13, which was on its way to Gibraltar. The eighty-two-ship convoy was also well supported by its own accompanying TF 64, with three destroyers, four destroyer escorts, and three Coast Guard cutters.[28]

The total "box score" for Captain Isbell's support group on this patrol was outstanding. Twelve good contacts were listed, with six submarines claimed sunk, five probably sunk, and one damaged. However, one pilot and one enlisted man were lost in the action, as well as one TBF and one F4F.[29] The damaged U-boat and four of the sinkings claimed by TG 21.14 have since been confirmed, and although HF/DF was not credited with contributing directly to any of the kills, the patrol resulted in valuable experience in its use.[30]

Annex C to Captain Isbell's report is a summary of search plans used. The general search plan was to approach any known submarine concentrations in such a way that a "daylight run" by the task group could be made straight through its center, without any change of course. Since the standard plane search radius was one hundred miles, this would enable the planes to keep the center of the concentration "under continuous observation throughout the entire day."

If new contacts developed, by HF/DF or whatever other means, the estimated center of the concentration could be adjusted accordingly. The task group would also make its night run so that it would be in position the next day to make the daylight run through the center of the latest estimated concentration.[31]

In an effort to take full advantage of the HF/DF equipment on board, several new, flexible air search plans were developed, and they were adjusted as necessary in response to HF/DF contacts. Plan 4, the "dawn search," was modified to provide for an initial short sweep ahead of the task group, "this being the resultant of our HF/DF ground wave contact North East of the Azores, at which time this initial short sweep ahead was not included and it is believed that we were sighted."[32]

It was apparent that HF/DF contact information was useful even when it did not lead directly to attacks on U-boats. Because HF/DF gave warning of the presence and location of submarines, it could be used to modify defensive procedures to conform to the current tactics of the U-boats. As Captain Brooks points out: "The primary consideration [in hunter-killer operations] was always making sure that the carrier was splashing through swept water."[33]

Annex C provides further information about the effective exploitation of HF/DF. The report continues: "When investigating a HF/DF bearing a scouting line was formed perpendicular to the line of bearing. If a definite fix had been established, a section of planes would proceed to the location and search, using the expanding square with the current visibility as the basic distance."[34]

The expanding square was one of many standard air search patterns. In addition to HF/DF contacts from his own equipment, Captain Isbell's daily summary of operations also noted information received from Washington. On 4 August at 1630, the *Card* catapulted two planes to search an area at 38 degrees north and 38.15 degrees west, "for sub indicated by HF/DF (COMINCH dispatch)."

The planes were aboard again at 2052, having seen nothing. On 5 August at 1731, the task-force commander broke radio silence and directed Lieutenant Commander Jones and Ensign Steward, who were already aloft, to search an area at 39 degrees north and 37 degrees west, "for a sub indicated by HF/DF (COMINCH dispatch). Results negative. Aboard 2049."[35] There was such respect for this information from headquarters that radio silence was broken in order to follow it up. Of course we now know that many of the commander in chief's dispatches gave information that had actually come from Ultra rather than from shore-based HF/DF, which was its cover.

Even on a patrol as successful as this one, the normal ratio of HF/DF contacts to actual sightings, and especially to kills, was very low. This was equally true, of course, for sonar and radar. It was one thing to receive a fleeting intimation of the possible presence of a U-boat, and quite another to actually locate it, prevent it from escaping, and destroy it.

On 12 August, for example, at 2225, the action report's daily resume noted a radar contact bearing 019 degrees at a distance of four thousand yards. At 2329 another radar contact was recorded, bearing 042 degrees, also at a distance of four thousand yards. "Neither contact identified," recorded the resume. The next day the old four-stacker *Borie* (which sank that November after being badly holed when rammed by a U-boat) reported a sound contact that was "shortly identified as false." These were typical daily entries in the resume of operations.[36]

Air operations against U-boats had a similar low ratio of sightings to attacks to kills. May 1943, which was the turning point in the Battle of the Atlantic, was a "record month" for British Coastal Command aircraft. But what this meant was that 213 sightings became 136 actual attacks, of which 17 were "considered to have been lethal"[37] and probably fewer actually were. Nevertheless, by 11 August planes from the *Card* had sunk the U-117, the U-664, and the auxiliary tanker U-525. Fido homing torpedoes had been responsible for two of the kills.[38]

By 23 August 1943 TG 21.14 had left convoy UGS 13 and was returning "en route to area of general concentration of outbound and homebound subs North and East of the Azores," accompanying the fifty-eight-ship convoy MGS 15. While close to the islands the TG could get air cover from land-based planes and did not need to conduct its own flight operations. There were two advantages to this: first, carrier operations were very costly in planes and men due to accidents, and second, land-based planes usually had access to a much larger supply of fuel.

The HF/DF contact reports continued for as long as TG 21.14 was in the Azores region, but they still yielded no definite results. On 23 August at 0900 the *Card* made an HF/DF contact, strength 4, bearing 024, which was described as "probably sighting report of this group." The contact was taken seriously, and at 0929 a plane was called off the port quarter patrol and directed to search for fifty miles along the HF/DF bearing. The "results," however, were "negative." On 25 August at 2332 an HF/DF sub contact was again reported, but this time "no exact bearing could be taken. Estimated distance over 60 miles."[39]

On 24 August aircraft from the *Card* sank another auxiliary tanker,

U-847, with Fido. This brought the total of U-boats sunk to four. By 6 September TG 21.14 was on its way back to Norfolk, making best possible speed. No flight operations were conducted during most of the return trip due to "heavy seas" and "hurricane weather." On 10 September TG 21.14 arrived safely back at base.[40]

At the end of the resume of operations for this patrol, Captain Isbell added a short summary of "new developments." Two items are of particular interest. On the subject of the daily sub estimates from the commander in chief, U.S. Fleet, estimates that had been arrived at using HF/DF, cryptanalysis, and other intelligence information, the report says that they proved "accurate and invaluable in our offensive search. No subs were sighted except in the locations given. The information regarding refueling operations [of U-boats] is always particularly positive and should be acted upon immediately."

And with regard to the HF/DF on board: "The new HF/DF equipment functioned satisfactorily, but without positive results. Three ground waves were intercepted, one apparently a sighting report Northeast of the Azores, but all developed negative. The two receivers furnished with this equipment are not sufficient to cover all probable frequencies. Four additional receivers have been requested."[41]

Unfortunately, although some action-report narratives distinguish what sort of HF/DF contact was obtained—ground or sky wave—the *Card*'s does not. The information about the three ground waves must have come from the operator's log. None of these logs is to be found with the action reports; indeed their whereabouts remain unknown.

It is also not known whether or not the *Card* obtained the additional receivers requested, though on its next patrol, between 25 September and 9 November 1943, its task group sank five more U-boats, again off the Azores. Captain Isbell's report for that patrol noted that "HF/DF has received only a few ground waves, all of which have resulted negative."[42] But that does not mean the device did not prove useful.

In the early hours of 1 November 1943 one of the *Card*'s escorts, the USS *Borie*, engaged U-405 in an hour-long contest involving torpedoes, gunfire, and rammings. Finally the U-boat sank, but the *Borie* had also sustained extensive damage and radioed the *Card* for assistance. Aircraft were sent out to the escort's last known position, but with negative results. Bad weather forced the planes back to the carrier early, before they had located the *Borie*. Nothing was heard between 0650, when the planes returned to the carrier, and 1110, when a message was received from the *Borie* that it had "commenced sinking."[43]

An HF/DF contact was obtained from the *Borie's* message, and at 1120 two planes were catapulted off to fly down the line of bearing toward the ship. At 1130 one of the planes located the sinking escort fourteen miles from the carrier, which changed course to go to the rescue. By 1630 the *Borie* had lost all power; it had fallen off into a trough and was taking on water. The order was given to abandon ship, and though this was accomplished in an orderly fashion, the bad weather and advancing darkness meant that of the 146 officers and men aboard, only 122 were rescued. Had it not been for the HF/DF bearing on the *Borie's* distress call, however, it is quite likely that the ship would not have been located in time to save anyone. By 0955 the next morning, 2 November, it was clear that there was no hope for the *Borie*, and the escort was sunk by a depth bomb.[44]

In the fall of 1943, following an HF/DF bearing was becoming increasingly dangerous for the escorts. By mid-September the suspicion was already growing among the Allies that a German return to the vital North Atlantic convoy routes was imminent. An increased number of U-boats had passed safely through the Bay of Biscay and were known to be moving west, codenamed group Leuthen. Decrypts had revealed that these boats had been especially prepared to strike back at the escorts that had driven them from the area in May. They were equipped with new devices such as Aphrodite, radar-deception balloons; and the Hagenuk radar search-receiver. They were also armed with a new weapon, the Zaunkönig acoustic homing torpedo, known to the Allies as the Gnat.

German tactics had changed, too, in response to the increased numbers and efficiency of the convoy escorts. The emphasis now was to go after the escorts first, with the homing torpedoes. The theory was that once the escorts had been knocked out or thrown into confusion, the merchant ships would be easy prey. On 20 September a series of signals to the Leuthen group from Dönitz in Offizier cipher was decrypted by the Allies. One of the signals read in part, "The decimation of the escort must be the first objective."[45] That very night the Leuthen group put the directive into practice, reopening the North Atlantic offensive by sinking one escort and damaging another.

There followed a three-day running battle pitting the Leuthen group of type VIIc U-boats against two nearby convoys and their escort groups.

The forty-ship convoy ON 202 and the twenty-seven-ship convoy ONS 18 ultimately merged, which gave them the clearly understood advantage of presenting a more compact and well-protected target. The British and Canadian escorts of group C2, with convoy ON 202, were two destroyers, a frigate, and three corvettes. They joined up with B3, which was escorting convoy ONS 18. B3 consisted of two British destroyers, a frigate, three corvettes, a trawler, and two Free French corvettes. These were reinforced by the Canadian Support Group 9, composed of the British frigate *Itchen*, the Canadian destroyer *St. Croix*, and three Canadian corvettes.[46]

This strong protective shield was fully utilized against the twenty U-boats participating in the attack. On 20 September a British frigate was badly damaged, and the *St. Croix* was sunk by a direct hit from a Zaunkönig while running down an HF/DF bearing. But early the next morning the U-229, located by HF/DF, was sunk by gunfire and ramming. Two more escorts were lost before the Leuthen group was called off on 23 September.[47]

The U-boats had been able to penetrate the escort screen to make just three attacks on the convoys themselves, sinking six merchant ships. In the seven multiple confrontations between U-boats and escorts, the escorts had generally held the wolves at bay. Extended periods of fog coverage and the U-boats' lack of radar greatly affected the outcome of the three-day struggle. Great support from land-based VLR Liberators from Newfoundland, as well as the eight Swordfish torpedo biplanes from the merchant aircraft carrier with ONS 18, made all the difference too; the Air Gap had ceased to exist.

But the greatest tactical advantage in this running engagement was without doubt conferred on the Allies by their use of HF/DF. Time and again between 20 and 23 September, HF/DF directed escorts to approaching U-boats and they were driven off or forced to submerge, often temporarily losing contact with the convoy as well as losing good attacking positions. The U-boats never suspected that their sighting and contact reports pulled the escorts down directly on top of them, and they continued to reveal themselves by their transmissions.[48] Indeed, statistics do not do justice to HF/DF, whose success is mostly measured in U-boat attacks not made—and therefore in ships saved.

By November 1943 the Allies' advances in technology, tactics, and experience enabled them to fight the convoys through even successive patrol lines. The U-boats continued to be hampered by lack of radar, while the Allies' electronic capabilities steadily improved. Radar-equipped air patrols and shipborne radar, sonar, and HF/DF combined to keep the U-boats submerged, preventing them from gathering into packs.

Once contact between the two forces had been made, what really defeated the wolf packs was the Allies' tactical superiority, given them by their electronic devices. By denying the U-boats surface mobility, the Allies seized and kept the tactical initiative. And because they remained unaware of the existence of shipborne HF/DF, the U-boats made the escorts' job of finding them easier by frequent transmissions. In December 1943 Dönitz disbanded his wolf packs, and no further large-scale operations were attempted in the Atlantic.[49]

From September to December 1943, the British and Canadians once again carried the main burden of fighting off the final assault of the wolf packs in the North Atlantic. Meanwhile, the U.S. Navy used much the same tactics in its continuing operations in the central Atlantic. The aggressive offensive procedures made possible by HF/DF increased the escorts' vulnerability, especially to the new Zaunkönig; nevertheless, they were better equipped than ever to carry out their essential function of securing the convoys' safe arrival.

The *Card*'s next patrol with TG 21.14, to the same general area off the Azores, lasted from 24 November 1943 until 2 January 1944. Probably because of sustained bad weather only one U-boat, the U-645, was sunk. By this time too, Captain Isbell seems to have become less confident about the accuracy of the commander in chief's daily U-boat estimates. These were being received several times a day, but now they often seemed stale by the time they arrived. On 23 December, for example, the sub estimate for the evening arrived at 1627. But in spite of the fact that it showed a concentration of around ten subs only eighty-five miles away from the task group, Isbell was reluctant to put too much faith in it because "previous estimates had often changed by the time we ran through them."[50]

Isbell added that the sub estimates could only be taken as a general indication of U-boats in the area because "we had actually run through the exact center of several active estimates already on this cruise during daylight hours without making any contact." But even if the estimate was accurate this time, and he actually was heading straight into a large sub concentration, Captain Isbell decided to pursue his direct course for Horta in the Azores because his escorts needed refueling urgently. The weather had been too bad to accomplish this at sea.[51]

But this time, the sub estimate was up to date. Captain Isbell's decision was to have a fatal result. At 2120 the *Card's* HF/DF reported the first of twelve B-bar enemy transmissions that were picked up within five hours that evening, all of them within a ten-mile radius of the carrier. The first contact, bearing 074 degrees, was followed two minutes later by a second bearing at 069 degrees, which was probably a weather report. Fifteen minutes later, at 2135, the *Card* got a radar surface contact. The old flush-deck destroyer *Schenck* was sent to investigate but the contact disappeared, submerging as the destroyer approached. However the *Schenck* made "a good sound contact" and attacked at 2143. The estimated results were that the U-boat had not been damaged.[52]

At this point Captain Isbell decided that he must protect his carrier, so the *Card* withdrew, taking evasive action at night and avoiding contacts. As he left Captain Isbell ordered two escorts, the *Schenck* and the *Leary*, to "keep subs down during night. We will be over in morning. Good luck."[53] Unfortunately the *Leary* was torpedoed while attacking one of the four U-boats that were in the vicinity. The old destroyer sank during the night, with the loss of almost one hundred men. Before the *Leary* went down, however, the *Schenck* had made two depth-charge attacks on a contact, the second of which destroyed U-645. It was Christmas Eve, 1943.[54]

Captain Isbell attributed the sinking of the *Leary* to its use of the TBS aircraft frequency and also to its use of a blinker in communicating that night because of the failure of the *Schenck's* TBS. At least something might be learned from this experience, as Isbell pointed out in his report:

> This action is believed to have proved that:
> a) Enemy subs remain submerged during daylight in areas covered by land-based aircraft.
> b) Enemy subs are afraid of our escorts and will not engage unless cornered.
> c) CVE TGs can cruise through enemy sub concentrations at night with reasonable safety.
> d) Star shells should not be used when radar is available.
> e) Enemy is equipped with radar.
> f) Enemy uses radar decoy balloons.
> g) More and better equipped escorts are required for CVE anti-submarine operations.[55]

With regard to HF/DF, the lessons to be learned were not quite so clear. Captain Isbell wrote: "Attention is again invited to the fact that all

HF/DF contacts except one were almost astern of this ship, although several radar contacts were on the flank. What if any significance should be attached to this fact seems open to question."[56]

What was not open to question was the crucial part played by HF/DF in the sinking of two U-boats by the *Block Island* and TG 21.16 two months later on the night of 29 February 1944. Capt. Logan C. Ramsay commanded the task group, which, in addition to the carrier, consisted of the destroyer USS *Corry* (recently used by the NRL to conduct preliminary HF/DF tests) and four new destroyer escorts, the *Bronstein*, *Breeman*, *Bostwick*, and *Thomas*.[57] As had been suggested by Captain Isbell in January, the number of escorts in escort-carrier task groups was increased to five whenever possible.

Initially the *Block Island*'s task group had been ordered out of Norfolk on 16 February 1944, in pursuit of a Japanese submarine believed to be en route to Europe. Failing to find any trace of the sub by 27 February, TG 21.16 was ordered to proceed to the "submarine concentration area" in the vicinity of fifty degrees north and twenty-five degrees west. This was the same general area in which the *Bogue* and the *Card* had been operating. The weather was dreadful for most of the cruise, often precluding flight operations. On 28 February, the weather finally cleared somewhat, and planes could take off.[58]

At 0300 on 29 February an HF/DF bearing was obtained on a three-group B-bar transmission, 3920 kilocycles. As a result of this contact, the task group changed course to 352 degrees to head in the direction of the bearing. Had this change in course not been made, the task group might never have located the U-boats.

Captain Ramsay believed the HF/DF contact "might have been a submarine. . . . In view of this possibility, confirmed by COMINCH dispatches indicating subs in the vicinity," he ordered the area searched by aircraft. Toward sunset a plane from the *Block Island* sighted a periscope at latitude 48°49' north, longitude 26°02' west. There followed a complex and drawn-out night action that continued into 1 March. "Two subs were believed to have been sunk and a third either sunk or badly damaged." The action was thought to have been against from five to seven subs, which fought it out at times on the surface and at times submerged. Both radar and sonar were used to locate the U-boats, and well-planned, coordinated tactics using radar-controlled gunfire, hedgehogs, and depth charges were all used in the attacks. As the report states, one of the subs was "probably sunk by a combination of gunfire hits by *Bronstein*, hedgehog hits by *Thomas,* and indeterminate depth charge damage by *Bostwick.*"[59]

It has since been confirmed that in the early hours of 1 March 1944, TG 21.16 sank the U-709 and the U-603.[60] HF/DF, at first a shipboard contact, later reinforced by general information from the commander in chief, put the task group in the right place to confront the U-boat concentration. And a successful mix of other electronic devices, tactics, and weapons secured a victory.

Later that month the experiences of TG 21.15 reemphasized the importance of training to the outcome of this highly technical war against the U-boat. On 24 March 1944 the task group, consisting of the escort carrier *Croatan* and five destroyer escorts, sailed from Hampton Roads along the North African convoy route via Bermuda, and then across the Atlantic to an area west of the Cape Verde Islands. On 7 April the U-856 was sunk, and the task group returned to base on 11 May. The action report made by the task-group commander, Capt. John P. W. Vest, was sharply critical of the lack of training of U.S. Navy pilots.[61]

Captain Vest was disturbed because in view of the current sub tactic of operating submerged during daylight and surfacing only after dark, it was imperative that pilots be trained in night flying. Yet Composite Squadron Forty-Two, aboard the *Croatan,* had "little total, and no recent night flying training."[62]

The lack of training in the use of specific equipment was even more distressing to Captain Vest.

> It does not appear that adequate provision has been made in the training program either of the aircraft squadron or of the escort group to insure efficiency in such combined operations. For example, the Escort Division assigned this TG had received no information whatever regarding sono-buoys and had never heard of the MK 24 mine (acoustical). The aircraft squadron assigned had never before used either equipment, had little instruction and no training in their use, and had received no instruction in capabilities and limitations of the DE sound equipment and search tactics.[63]

Appended to this action report is a special "Study of Operations" drawn up on 22 May by Captain Heineman, then of the Atlantic Fleet Anti-Submarine Warfare Unit. HF/DF comes off well as compared with other deficiencies in training. Under point sixteen of the study, "Employment of special devices, HF/DF," Captain Heineman stated, "On at least one occasion this device again proved helpful in locating the enemy and this TG demonstrated ability in its employment."[64]

Bob Seroskie began his duties as HF/DF operator at about this time. In a recent letter he summed up the use to which his training was put:

> We never did pick up any signals from a sub transmitting in our vicinity. My experience with Huff Duff spanned the period from about March 1944 to May 1945. The fact that we operated off the Northeast and Florida coasts with a carrier and in a declining period of sub activity are factors in our limited role with the equipment. I do however believe that Huff Duff would have been quite productive had it been available during the early stages of the war. I say this only on the basis of my confidence in the capabilities of the equipment."[65]

Even in the summer of 1944, though, there was still work for HF/DF in some areas. One of the most spectacular events in the U.S. Navy's war against the U-boats in the Atlantic was the ramming and sinking of the U-66 by the destroyer escort *Buckley*.

The *Buckley* was sailing with the *Block Island* (which would be sunk by a German torpedo on 29 May) in TG 21.11, departing Norfolk on 22 April 1944 to operate against U-boats west of the Cape Verde Islands. At 0220 on 6 May the *Buckley*, under Lt. Comdr. Brent M. Abel, was ordered to approach and attack a surfaced sub that had been spotted by a plane about eighteen miles away. At 0221 "DAQ picked up 6 group B-bar transmission on 4790 kc, on borderline between sky wave and ground wave. No bearing established." At 0240, however, "DAQ picked up a repetition of Enigma bearing 317 degrees (directly on the bearing of the sub, which had been also established by radar fix)." Following that bearing, and directed by the plane that was still overhead, the *Buckley* suddenly loomed up on the U-66, which had remained on the surface.[66]

After a fierce running battle in which gunfire was exchanged, the *Buckley* rammed the U-66, and for a short time the two vessels were locked together. Desperately scrambling off their badly damaged submarine, the Germans tried to save themselves by climbing onto the destroyer that was wedged alongside. Under the impression that he was being attacked, Lieutenant Commander Abel is reputed to have given the historic, and long unused, naval command, "Stand by to repel boarders."

For a while, a melee took place with the Americans using any weapon that came to hand, including spent shell casings and at least one coffee mug, to fight off the "attackers." But as the sub broke away and sank the fighting ceased, and the *Buckley* picked up thirty-six survivors.[67]

One of them was Petty Officer Karl Degener-Böning, the sub's radio operator. He remembers that by 6 May the U-66 had been at sea for nearly sixteen weeks without resupply.

> In the night from 25 to 26 April, we were to meet U-488 for refueling and provisions. But the moment we sighted the boat, she was attacked and sunk by destroyers. We heard it. From that day on we tried to get out. In daylight we had to dive. Once the *Block Island* was sighted by periscope, but the carrier was too far away [to attack], and all the time planes were in the air. At night we had to come up to load the batteries and to get fresh air. Often we were forced to dive again by planes. On 5 May, finally, our commander, Seehausen, decided to stay on the surface to give detailed information about our situation to HQ. He planned to reach the Cape Verde Islands. The end, as you know, was the clash with the *Buckley*.[68]

This clash, fatal to the U-66, was brought about at least in part by the DAQ contact that helped to establish the location of the submarine. The episode is also notable because, as a matter of record, twenty-six separate HF/DF contacts were registered on that one final high-frequency message transmitted by Karl Degener-Böning on Captain Seehausen's orders. Those orders were mandated by Admiral Dönitz's insistence on being kept up to date on the situation of each of his U-boats.

It is perhaps fitting to let Karl Degener-Böning have the last words about this engagement: "The decision by [Lt.] Comdr. Brent Abel to ram us, instead of fighting us down with other weapons such as depth bombs, saved the lives of 36 men of our crew of 60 men. His gallant behavior after the fight, to order our rescue, was an outstanding action."[69]

HF/DF had been designed, built, installed, and manned to intercept German U-boat transmissions. It was used, just as planned, in the location and sinking of the U-66. But another operation illustrates, as did the case of the *Borie*, that HF/DF could be used in a surprising number of different ways. On 24 July 1944, the USS *Bogue*, with Capt. Aurelius B. Vosseller in command of TG 22.3, departed Norfolk to relieve the TG 22.6 with the *Wake Island*.

TG 22.6 had been deployed against the weather U-boats known by radio interception to be operating in the North Atlantic. In addition to the escort carrier and four destroyer escorts, six Canadian frigates joined

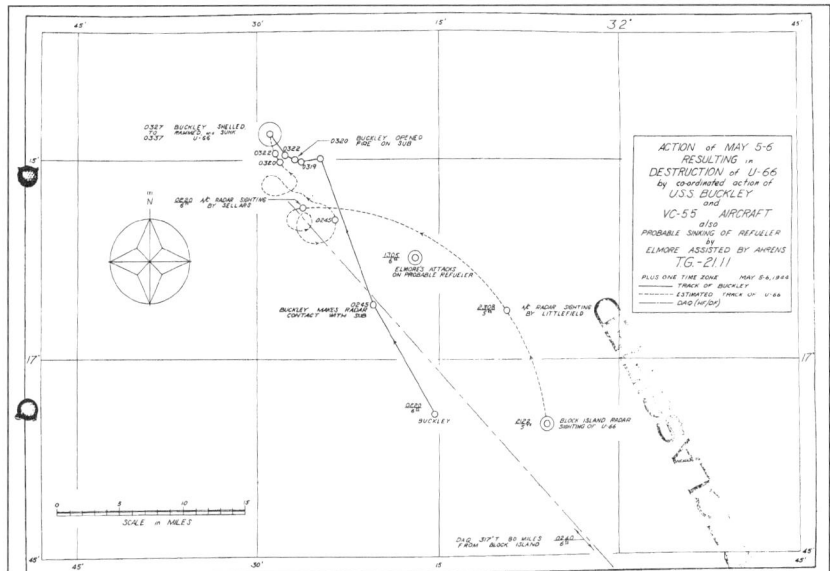

Map 9.1. Report of antisubmarine operations by TG 21.11, 22 April–29 May 1944. *NHC/OA, Action Report, Box 103*

the *Bogue* in August. On 20 August, six aircraft from the *Bogue* sank the U-1229 on its way to land secret agents on Long Island. On 16 September a TBM piloted by Ensign Schroeder was carried out of position by very high winds. After patrolling for six hours and twenty minutes, Ensign Schroeder was forced to land in the water when he ran out of fuel.[70]

Long before ditching, Schroeder had known he was lost, but the *Bogue*'s SK radar was unable to pick up the plane while it was further than fifty miles out. Using HF/DF bearings obtained by the destroyer escort USS *Willis,* "the plane was brought into SK range and was vectored to within twenty miles of the ship before fuel was exhausted . . . [Ensign Schroeder] and his crew were rescued uninjured!"[71]

There are many other accounts of operations involving HF/DF's use by the Royal Navy and the RCN, as well as by the U.S. Navy. Ultra and shore-based HF/DF were used for the diversion of convoys as well as to direct support and hunter-killer groups to the general location of gathering U-boats. Shipborne HF/DF was of equal importance to the tactical success of the U-boat war. As well as being used for the tactical diversion of convoys and the rescue of pilots and seamen, shipborne HF/DF (and radar and sonar) was irreplaceable for the location and destruction of U-boats. But it is almost impossible to determine which factor was "decisive" in each particular case.

Not only were there too many variables involved in each action, but the record that remains is spotty and only as good as the men who wrote the reports. Even in a peacetime exercise, few of those men would have been qualified or indeed able to judge which of the many devices brought into play could actually be called decisive. In the heat of battle there was tremendous confusion, and many errors must have been made, many details forgotten, by the time the action was written down.

An action report pad was kept on the bridge of each U.S. Navy vessel specifically to record all significant activity. At the top of the pad was the helpful reminder, "Attack First, then collect data for this report. See back of pad for instructions."[72] Indeed, reporting was so often inaccurate and fragmentary that the Operations Evaluation Group was constantly frustrated by the flawed operational data they had to use.[73]

The part played by HF/DF in defeating the U-boats in the Atlantic was a broad, general one that, like the part played by radar and sonar, is hard to describe quantitatively. The number of U-boat kills to which the use of HF/DF can be proved to have been directly related is only one element in the story of its overall effectiveness. While it may be one of the most exciting elements, it is probably not the most important. The best estimate of U-boat sinkings verified by the U.S. Navy shows that twenty-two German submarines were sunk between 22 May 1943 and 8 April 1945 in ASW attacks aided by HF/DF.[74] In addition, one U-boat was captured (the U-505, on 4 June 1944), and several others were damaged. This compares with fifty-four sinkings in which Ultra information was used, according to one analysis.[75]

If the total number of U-boats sunk by all means during the war is estimated at about 733, then HF/DF has been shown to have been involved in about 3 percent and Ultra in about 7 percent of those actions.[76] However, forty-two German submarines were sunk by the U.S. Navy hunter-killer groups in 1943 and 1944.[77] Two of the HF/DF-aided sinkings on the U.S. Navy list were Royal Navy kills and one was credited to land-based air from San Juan, Puerto Rico. The remaining nineteen sinkings aided by HF/DF and credited to the U.S. Navy represent 45 percent of the forty-two confirmed sinkings. Not a bad record.

But U-boat kill numbers may not be the best measure of effective-

TABLE 9.1
ASW ATTACKS AIDED BY HF/DF

Date of use of HF/DF	Units involved	Results
29 Feb. 1944	TG 21.16 (*Block Island, Thomas, Bostwick, Breeman & Bronstein*)	U-358, U-709 & U-603 One probably damaged
11 Aug. 1943	*Card* (2 Avengers)	U-525 sunk
6 Nov. 1943	HMS *Tracker's* group	U-842 sunk
23 Dec. 1943	*Card* group (*Schenck, Leary & Decatur*)	U-645 sunk
7 Aug. 1943	San Juan, P.R. (A/C & S/C) *	U-615 sunk
17 May 1943	*Moffett & Jouett*	U-128 sunk
26 Apr. 1943	*Croatan* group (*Snowden, Frost, Inch & Barber*)	U-488 sunk
5 May 1944	*Block Island* group (*Buckley, Ahrens & Eugene E. Elmore*)	U-66 sunk
4 June 1944	*Guadalcanal* group (*Pillsbury & Chatelain*)	U-505 captured
30 Sept. 1944	*Tripoli & Mission Bay* groups (10th Fleet trackers)	U-1062 sunk & U-219 not damaged
15 June 1944	*Croatan* group	U-853 sunk
13 Aug. 1944	*Bogue* group	U-802 no damage
16 Jan. 1945	*Otter, Hubbard, Varian & Hatter*	U-248 sunk
15 Apr. 1945	*Croatan* group (*Stanton & Frost*)	U-1235 sunk
15 Apr. 1945	*Croatan* group (*Stanton & Frost*)	U-880 sunk
16 Apr. 1945	*Mission Bay* group (*Mosley*)	U-805 no damage
21 Apr. 1945	*Croatan* group (4 escorts)	U-805 no damage
21 Apr. 1945	*Croatan* group (*Carter & Neal A. Scott*)	U-518 sunk
24 Apr. 1945	*Bogue* group & escorts	U-546 sunk
8 Apr. 1945	*Mission Bay* group (*Farquhar*)	U-881 sunk
30 June 1942	VP-74 A/C	U-158 sunk
14 July 1943	*Santee* group	Forced U/B to submerge.
14 July 1943	*Santee* group (*Bainbridge*)	U-160 no evidence of damage
14 July 1943	*Santee* group (A/C)	No evidence of damage
14 July 1943	*Santee* group (A/C)	No evidence of damage
15 July 1943	*Santee* group (A/C)	U-509 oil slick on surface
14 July 1943	*Santee* group (A/C)	Small amount of oil
24 July 1943	*Santee* group (A/C)	Small moving oil slick
30 July 1943	*Santee* group (A/C)	Two U/Bs attacked; one left a small oil slick and U-43, the second boat, left a small amount of debris and oil.
22 May 1943	*Bogue* group A/C	U-569 sunk
23 May 1943	HMS *Keppel & Archer* A/C	U-752 sunk
4–9 June 1943	*Bogue* group A/C	Attacked 3 U/Bs but failed to sink them.
18 June 1943	*Bogue* group A/C	U-118 sunk

TABLE 9.1 *(CONTINUED)*

Date of use of HF/DF	Units involved	Results
21 Aug. 1943	*Croatan* group A/C	U-134 no damage
21 Sept. 1943	ONS 182, ON 202 with combined escorts	Group Leuthen
21 Sept. 1943	HMCS *Chambly*	U-584 slight damage
22 Sept. 1943	British trawler *Northern Foam*	U-852 scored a few hits
22 Sept. 1943	HMS *Keppel*	U-229 sunk
22 Sept. 1943	Newfoundland-based A/C	U-270 damaged
22 Sept. 1943	Newfoundland-based A/C	U-377 damaged
22 Sept. 1943	Newfoundland-based *Liberator*	Engaged U-402. Expended all bombs and was driven off by U/B's AA guns.
22–23 Sept. 1943	*Itchen*	Held the U/B down
23 Sept. 1943	SS *James Smith* armed guard	Scored hits on C/T of U-666
7 Oct. 1943	Polish DD *Orkan*, HMS *Musketeer*, *Oribi* & *Orwell*	U-758 & U-378 no damage, but *Orkan* was sunk.

Source: NHC/OA Ready Reference Files: HF/DF
* A/C = aircraft; S/C = subchaser

TABLE 9.2
CONVOY DIVERSIONS AIDED BY HF/DF

Date	Convoy
28 Apr. 1943	ONS 5
4 June 1943	GUS 7A
7 June 1943	UGS 9
9 June 1943	UGS 9
21 June 1943	GUS 8
18 Sept. 1943	ON 202
23 Sept. 1943	ONS 18
9 Dec. 1943	GUS 23

Source: NHC/OA Ready Reference Files: HF/DF

ness of HF/DF. A better measure, but one even harder to estimate, would be the number of merchant ships saved or the additional cargoes safely delivered.

Even if the quantitative results indicated for tactical offensive use of HF/DF accounted for only a small portion of U-boats sunk, the Atlantic War was so close run that every small advantage made a real difference. Huff Duff may not have been the decisive weapon against the German U-boats, but it certainly accelerated the Allied victory. If one ultimate measure of effectiveness for World War II is the date on which Germany finally surrendered, then HF/DF made a significant difference by helping to cut short the U-boat campaign.

Clearly the timing of events in the Atlantic was crucial to the whole war, and Henri Busignies's Huff Duff affected that timing at two pivotal points. The first was in January 1941, when Busignies appeared in the United States with the design for an efficient land-based direction finder just when it was needed. The second occurred in March 1942, when Busignies's ability to immediately supply an effective shipborne HF/DF encouraged the U.S. Navy command to go on the offensive against the U-boats.

CHAPTER 10

The Verdict on Huff Duff: World War II and After

How much had Dönitz's attempt to starve Britain into surrender cost Germany? During the war, 1,170 U-boats had been commissioned. Of these, 860 made at least one patrol and 630 of them were lost, or approximately 73 percent. An additional 81 boats were lost in the home area and in harbors from mines and air attacks, and 42 were lost in those areas to accidents. So 753 boats were lost outright, and 38 more were decommissioned during the war because of irreparable damage. In human terms, the toll was devastating. Of the approximately 41,000 Germans engaged in operations, 27,378 were killed (including Dönitz's two sons), and 4,945 became prisoners of war.[1]

Among the Allies the losses were even heavier. The total of Allied and neutral merchant ships lost to submarines was 2,828, and of the 5,150 global total of vessels lost, 2,452 went down in the Atlantic. This totaled 12.8 million tons.

It is impossible to know exactly how many people died in the campaign, though the British merchant navy alone lost 30,248 men, most of them in the Atlantic. U-boats in the Atlantic also sank 175 Allied warships, most of them British, and the Royal Navy lost 73,642 men, the majority in the Atlantic. The Royal Canadian Navy lost 1,965 men.[2] The U.S. Navy lost 36,950 men, mostly in the Pacific. In the Atlantic the losses in American men were somewhat under 2,000. U.S. merchant mariners suffered 6,833 dead.[3] Each of the Allied air forces in the Atlantic lost planes and men as well, the British numbering losses of both in the thousands.[4]

This terrible campaign was fought with unflagging determination by both sides, and the margin of victory was narrow. HF/DF was one of the important elements in that victory. Dönitz designed a system for the command and control of his U-boat force that relied heavily on two-way radio transmissions. It was the accessibility to detection of this system, successfully exploited by the Allied naval forces, that gave to Huff Duff such a crucial role in the Battle of the Atlantic.

And still, almost to the end of the war, Dönitz continued to require his U-boats to transmit. He was never able to "see" the open secret of Huff Duff. He was not able to accept the evidence of Allied use of HF/DF against him. Perhaps this blindness was a form of institutionalized arrogance; Dönitz may not have been able to acknowledge that the fatal weakness of his U-boats lay in his own system of radio control.

Often when he radioed a U-boat at sea, there does not seem to be any obvious reason why he could not have supplied the information in sealed packets, along with the rest of the instructions each skipper received before sailing. And Dönitz's men apparently did not share the traditional U.S. Navy aversion to "rudder commands from the beach."[5]

When targets disappeared or U-boat losses mounted, Dönitz and his staff looked for spies to blame or secret enemy technical devices, such as an all-powerful radar. Perhaps HF/DF was too obvious. Apparently Dönitz was so wedded to the radio that in spite of periodic concern with the danger of detection, in his mind this never outweighed the advantages of close personal control.

And even if Dönitz's naval intelligence and scientific advisers understood this danger themselves, they were never influential enough to secure a change in policy. Particularly in the case of the failure to guard against Allied direction finding, it is clear that the German scientific establishment was not as well integrated into Dönitz's operational headquarters as it should have been.

It was fortunate for the Allies that it was not. The constant use of radio provided them with a weak link they could attack, and they did so quite effectively, not only with the strategic use of information from decryption, but also with the tactical use of shipborne direction finding. In the escalating war of weapons and electronic devices, where every new measure was quickly matched by a countermeasure, Huff Duff had a unique weakness: it could not work if the U-boats did not use their radios. Of all the electronic devices marshaled against Dönitz's U-boats, HF/DF was the most easily countered.

Dönitz had at hand the perfect foil to Huff Duff: stop transmissions from his submarines to headquarters. He would still have been able to exercise a great deal of control over his boats. Being assured of the security of his Enigma ciphers, he would still have been free to send his U-boats coded instructions. He could have continued to direct them onto targets by transmitting the B-Dienst decrypts of Allied convoy routing instructions. But he chose to insist that his U-boats also send him data in return. As a result, the frequent radio bulletins flying along the airwaves from U-boats to headquarters were like arrows pointing right back to the spot from which they had come. The British and the Canadians each developed their own responses to the German submarines in the Atlantic, and in many ways the ultimate outcome of the campaign rested with them. The focus here has been on the American response. In particular the issue has been how the U.S. Navy acquired an electronic device that could turn Dönitz's use of the radio against him.

The provenance and subsequent development of American Huff Duff provides insight into the navy's adoption of a needed technology in wartime. HF/DF is more truly an American story than previously appreciated, because the HF/DF devices deployed by the U.S. Navy were not British in origin as widely believed, but rather were the products of Henri Busignies and ITT. HF/DF came to be identified almost haphazardly—it was developed by essentially unfettered private industry—but it was deployed and used effectively against German U-boats by the navy.

The case of HF/DF shows the decision makers in the U.S. Navy and related civilian groups to have been varied, numerous, and only roughly coordinated. The navy, administrative and operational, had worked with scientists and industry to achieve victory in the Atlantic. But cooperation had not always been easy. The postwar report of the Naval Research Laboratory candidly admits that "the war forced the naval scientists to take the engineers and scientists of the industrial and academic world into their confidence."[6]

Often the bureaucratic efforts, both military and civilian, to direct and control wartime scientific and industrial development had been less than effective. In a classic reversal of the normal top-down approach in which a weapon or device is designed to meet specific requirements, it was the entrepreneurial spirit of Sosthenes Behn that promoted HF/DF, even before the navy was aware of its usefulness. Finally, it was the tardy recognition of the need for such a device in the navy that circumvented all bureaucratic objections. These objections included seriously entertained—though patently foolish—fears about security. In fact, one paradox illustrated by the development of HF/DF is that there was often a reduction in the observance, if not the existence, of red tape during the war. This occurred in spite of the proliferation of control mechanisms such as the National Defense Research Committee, the Office of Scientific Research and Development, the War Production Board, and others for planning, oversight, and inspection.

In other words, those who made decisions with regard to what technologies the navy required for the prosecution of the war were generally able to secure that technology. This was true even when, as with HF/DF, it meant sidestepping serious attempts by the Office of Naval Intelligence to restrict the activities of ITT in the name of national security. Henri Busignies's Huff Duff was developed in spite of the qualms of the intelligence establishment.

This suggests that extreme security measures may not work well in every case in wartime and may even be counterproductive. Radar was a tightly guarded Allied secret, yet each advance in its development was quickly countered by German scientists. On the other hand the open secret of HF/DF, though a persistent possibility, was never identified by the Germans as a real danger. Even the significance of its telltale, highly visible, and distinctive antennas was ignored.

World War II was the first truly electronic war, and the flexibility of an economy fueled by competitive private enterprise made the United States particularly suited to dominate this relatively new field. There was a vast increase in the amount, the quality, and the diversity of electronic equipment produced in the United States during the course of the war. This was especially necessary for the navy, a technical service, and one that was flooded with wartime recruits lacking in sea skills. The consequence was an ever-increasing reliance on technical aids.

In order to minimize training time for these new devices, there was a premium on simplifying technological equipment. This was a large part of the significance of Busignies's automatic, direct-reading, instantaneous, HF/DF device. The DAQ was easier to operate than the equivalent British FH3, and it gave comparable or better results.

Military systems appear to be particularly prone to fads, to the search for panaceas—the perfect war-winning weapon, vessel, or device. Even in the electronic war, however, Huff Duff was not that device. Nor was sonar, nor radar, nor even, in spite of its tremendous significance, Ultra. The truth is that on its own Huff Duff could not be very effective, nor could any other device. They were only significant as they formed part of an integrated operational system. It took a shifting, ever-improving kaleidoscope of weapons, vessels, and electronic devices, used in cooperation with each other by experienced forces, to defeat Dönitz's skilled and dedicated submariners.

The effectiveness of men and equipment was also dependent on being wedded to an appropriate doctrine. When the U.S. Navy was predominantly on the defensive against U-boats in the Atlantic, shore-based HF/DF provided valuable information that helped to route convoys away from danger areas. But the information on U-boat locations derived from shore-based HF/DF was not available to escorts immediately enough to be useful in tactical operations. When the U.S. Navy went on the offensive in the Atlantic in 1943, it was absolutely essential to have shipborne HF/DF if the device was to play a major part in obtaining U-boat kills.

Immediate postwar revelations gave credit to the role of American HF/DF in the Atlantic. Finally freed from wartime security constraints, the publicity was almost embarrassing in the extent of its claims. On 14 January 1946 a front-page headline of the *New York Times* read: "'Huff-Duff,' U-Boat Finder, Seen As Defense Against Atomic Bomb." The *Times* claimed that "no radio signal in space seems to elude this master sleuth," and that if war were to come again Huff Duff could "cope with an enemy's atomic bombs shot into space thousands of miles away."[7]

This article was the result of a meeting on 12 January, when the navy, in cooperation with engineers from Federal Radio andTelephone Company, first disclosed the story of Huff Duff to the press. They held a series of demonstrations for scientists as well as for journalists at the ITT experimental radio laboratory at Great River to illustrate how the technology worked. The *Times* reported that according to the naval officers present, Huff Duff was used during the war to "track and locate the hordes of Axis submarines in the Atlantic." It accomplished this task so successfully that "sinkings of Allied shipping due to submarines in the first few months of the use of Huff Duff were reduced by a factor of ten to one, and finally by a fifty-to-one ratio."[8]

The navy spokesmen at the demonstrations credited the conception of Huff Duff to Dr. Henri Busignies, from his prewar days at the Paris laboratories of the International Telephone and Telegraph Company. Then, it was explained, Dr. Busignies had moved to the United States early in the war and become director of the Federal Telecommunications Laboratories, an affiliate of ITT. The journalists were told that at Federal, along with divisional head in charge of direction-finder apparatus A. G. Richardson, Busignies had developed his Huff Duff designs "in cooperation with the armed forces."[9]

The information revealed about World War II American Huff Duff at these briefings was both accurate and detailed. It is surprising that subsequent historical accounts have been so skimpy and imprecise. On 14 January, following the publicity demonstrations at the Great River labs, many newspapers around the country ran extensive articles on Huff Duff. Most of these gave a very clear, scientific explanation of the operation of the device, including the facts that it scanned incoming waves from all directions and distances, that it measured the angle of arrival with reference to true north, and that it pictured the result in terms of a dot or streak of light on the face of a cathode-ray screen.[10]

Part of what made the Great River presentation so memorable for the guests was a practical demonstration of the speed and accuracy of Huff Duff. A bearing for a particular transmitter on Bermuda was requested by telephone from the Atlantic coast control station, which was operated by the Army Airway Communication Service at nearby Mitchell Field. The same bearing was requested from various of the former navy receiving stations, which had recently been handed over to the Coast Guard.

In less than a minute the bearings on the Bermuda transmitter were back at Great River. In less than two minutes they had been plotted on a huge map set on a table. The map was made of a thin sheet of aluminum to avoid inaccuracies due to paper shrinkage. A marker was pressed in where the bearings crossed, and this proved to be less than ten miles on the chart from the actual transmission station on Bermuda. This would have put the transmitter within eyesight of a destroyer sent to the area, had the transmission come from an enemy interloper.[11]

Navy and Federal officials also explained the ongoing postwar importance of Huff Duff, pointing out that six receiving stations were constantly on guard from Maine to Florida. These stations were linked by high-speed telegraph and teletype, so they could operate cooperatively in emergencies.

Other stations were operating at various airfields on the Atlantic and Pacific coasts, and in Africa, Europe, and the Far East. In fact, more than a thousand land-based HF/DF receiving stations worldwide, as well as hundreds more on ships, were ready for use in emergencies. Their task was to help protect life and property at sea and in the air by locating ships or planes that were lost or in distress. It was claimed that "hundreds" of these emergencies were being handled monthly.[12]

But apart from the peaceful use to which Huff Duff could be put, its great range and accuracy also lent itself to another purpose, which navy spokesmen at Great River made uncompromisingly clear. If Berlin were to direct radio-controlled V-bombs at Times Square in another war, "we would soon know it by high-frequency direction finders. We could locate the bombs in the stratosphere, explode them, or turn them aside."[13] Berlin was still the projected enemy, not Moscow, and there was as yet no thought of the need to locate Russian submarines. Postwar Huff Duff was to be both the savior of lost travelers and the magic electronic shield protecting the United States from enemy rocket attack. HF/DF was seen as the Star Wars of the late 1940s.

Like sonar and radar, Huff Duff promised to have a continuing role to play after World War II. Yet unlike its electronic cousins, *Huff Duff* never became a household word. This is partly because of postwar developments in pulse-navigation systems such as Loran and Tacan, which made them better for navigational purposes than the direction finder. Loran and Tacan have become well known to boating and air enthusiasts and are standard equipment on all sorts of pleasure craft. High-frequency direction finders are familiar mostly to professionals in the field of electronics.[14]

It has been suggested that the real power of direction finding has been intentionally obscured because of its continuing significance for national security. According to this line of thought, in order to encourage the Russians to use radio communications, the Americans hid their very potent capacity for direction finding off Russian transmissions.

In 1977 the U.S. Navy sponsored a secret study of many aspects of antisubmarine warfare from 1940 onward. This included an investigation of HF/DF. The resulting Cross report, completed in 1978, was based almost entirely on classified material and was not itself declassified until December 1990.

Apparently HF/DF remained a current issue in ASW after World War II, at times more important than radar. The Cross report quotes the 1954 testimony of a navy captain in the Anti-Submarine Plans and Policies Group to the effect that "radar is going downhill with respect to effectiveness and eventually must be out of the picture with the true submarine."[15] The postwar "true" submarine, which could remain submerged almost indefinitely, was impervious to radar. This made underwater sound surveillance systems more popular than ever, and it bred sophisticated new systems like Lofar (low-frequency analyzing recorder).

And a report from the May 1959 U.S. Navy Anti-Submarine Conference lists as one of six essential requirements the "provision of HF/DF capability for intercept of time-compressed signals from Soviet subs."[16] The Cross report in fact outlines extensive postwar developments in navy direction finding. There certainly was an ongoing postwar concern with HF/DF, which made public discussion of its capabilities a security issue.

From the opposite perspective, the debate about the wisdom of long-range radio control of naval forces still continues too. In 1989 the National Research Council and the Office of Naval Research conducted a study at the request of the chief of naval operations. The study "looked into the implications of advancing technology for naval operations in the twenty-first century."[17] Known as *Navy 21*, this analysis proposed a "new, fully integrated radio-electronic battle management (REBM) system encompassing surveillance, combat information, and command, control and communications 'as an organic whole.'"[18]

Navy 21 contended that such a system "would improve the operational commander's ability to target enemy forces."[19] Fifty years earlier Dönitz might have used those same words to describe his own system of radio control of his U-boats. The study added that REBM would also prevent the enemy from targeting his own forces. But Dönitz too relied on the difficulty of tracing his high-frequency signals to prevent his U-boats from being targeted by their use of radio.

Not everyone involved in *Navy 21* supported its conclusions. One participant, Vice Adm. John Boyes, doubted that even a twenty-first century REBM system could "close the crucial gap between the tactical action officer and the commanding officer."[20] Proponents of REBM cited the virtues of a highly centralized command structure. But others, "especially those with World War II command experience, such as Adm. Arleigh Burke and Gen. John Vessey, argue for decentralization and dispute that a battle can be managed from afar."[21] Perhaps they had in mind Admiral Dönitz's ill-fated attempt to do just that.

How important had Huff Duff really been to the Allied victory against the U-boats? While he had command of the *Belknap*, Ben Brooks's total "bag," using all weapons, and with the advantage of radar, sonar, and Huff Duff, was one whale and, he unhappily believes, one friendly French submarine. But Ben Brooks's war was only one small part of a vast, lengthy, and constantly changing campaign.[22]

Most historians agree that success in the longest battle of World War II was the product of a multiplicity of factors, each playing an important role, each a part of a larger, complex whole. Convoying; escort carrier groups; air coverage; new weapons and tactics; electronics such as sonar, radar, and HF/DF; especially Ultra; and American industrial output have each taken a turn in the spotlight.

In terms of intelligence, what Ultra supplied to the Allied forces afloat in strategic information, HF/DF supplied in tactical information. As a tactical device for the location of U-boats, shipborne HF/DF was constantly in use in the Atlantic from mid-1942 onward, picking out U-boats for attack every time they went on the air. Even more than radar or sonar, HF/DF deprived the U-boats of their stealth.

Clearly, Henri Busignies and his Huff Duff devices added significantly to the American contribution to success in the Battle of the Atlantic, and hence to ultimate victory in World War II. This was true in spite of the fact that direction finding was a well-known technology studied by scientists in many different countries—including Germany—in the years between world wars. Admiral Dönitz had even anticipated the use of HF/DF against his radio-communication system. But he failed to recognize the appearance of a greatly improved device, one that could actually be used as an effective U-boat locator. While HF/DF itself was no secret, it did become an invaluable secret weapon.

Notes

Preface

1. Morison, vol. 1, *Battle of the Atlantic*, 226–27.

2. Hezlet, *Electronics*, 230.

3. See, for example, David Syrett, "The Safe and Timely Arrival of Convoy SC 130"; "Weather-Reporting U-Boats in the Atlantic, 1944–1945"; and especially his recent book, *The Defeat of the German U-Boats*. See also Jürgen Rohwer's classic work *The Critical Convoy Battles of March 1943*.

4. Till, "The Battle of the Atlantic," 584.

Chapter 1   An Overview

1. Plante, "A Very Personal View." Memoir, 1991.

2. See Middlebrook, *Convoy*, for a very detailed account of convoys HX 229 and SC 122. Total merchant tonnage lost was 146,596 tons. See also Rohwer, *Axis Submarine Successes*, 157–60; and Rohwer and Hummelchen, *Chronology*, 200–201.

3. Hackmann, *Seek and Strike*, 238.

4. Rohwer, "The Wireless War," 408.

5. Morison, vol. 1, *Battle of the Atlantic*, 226–27. This is the classic view. There are a few errors in Morison's account of HF/DF, which have, unfortunately, been perpetuated by his followers. On p. 228 Morison states that "Dr. Harry Goldstein was largely responsible for the American version [of HF/DF]." He means, of course, Dr. Maxwell K. Goldstein, wartime head of the Direction Finder section of the Naval Research Laboratory. However, it was not Dr. Goldstein at all, but French engineer Dr. Henri Busignies who was responsible for creating the American HF/DF. This was not a version based on borrowed British Admiralty sets, as implied in Morison; it was uniquely his own.

6. Rhys-Jones, "The German System," 138.

7. Hessler, *U-Boat War*, vol. 1, 40–45, 86.

8. Erskine, "U-Boats," 326. See also Rhys-Jones, "The German System," 138, 142–47.

9. Keegan, *Price of Admiralty*, 266.

10. Barnett, *Engage the Enemy*, 575–76.

11. Kahn, *Seizing the Enigma*, 5–6. The quote is from Keegan, *Price of Admiralty*, 266. Import and loss figures vary considerably in different accounts.

12. Plante, memoir.

13. Churchill, *Their Finest Hour*, 598–99. See also Erskine, "U-Boats," 326.

14. Middlebrook, *Convoy*, 12. See also Lundeberg, "Allied Cooperation," 354–55.

15. Dönitz, *Memoirs*, 115–16.

16. Dönitz, *Memoirs*, 63. Emphasis is Dönitz's own.

17. For Ultra see Kahn, *Seizing the Enigma*; Winterbotham, *The Ultra Secret*; Smith, *The Ultra-Magic Deals*; Beesly, *Very Special Intelligence*; and Syrett, "Communications Intelligence."

18. For the role of HF/DF, see Fisher, *Race on the Edge of Time*, 309; and van der Vat, *Atlantic Campaign*, 176. For the most recent and thorough study of tactical use of HF/DF, see Syrett, *Defeat of German U-Boats*; and "Spot Location by HF/DF Data," *Aero Digest*, 63.

19. Watson-Watt, *Three Steps*, 360.

20. Hessler, *U-Boat War*, vol. 1, 77–80.

21. van der Vat, *Atlantic Campaign*, 55, 116–17, 142–43.

22. Dönitz, *Memoirs*, 12–14. See also Terraine, *U-Boat Wars*, 290.

23. Howse, *Radar at Sea*, chaps. 1–3.

24. Richardson, "Galena," 2.

25. See Kahn, *Seizing the Enigma*, 82–103, 185, for some of the denizens of Bletchley Park of whom Churchill was said to have remarked, "I know I told you to leave no stone unturned to find the necessary staff, but I didn't mean you to take me so literally."

26. Friedman, *Submarine Design and Development*, 48 (hereafter *Sub D & D*).

27. There was no airborne HF/DF. While some aircraft were fitted with low-frequency direction finders, they were for navigation and were of no use for hunting U-boats transmitting on short waves. Aviation technology in general will not be dealt with here. See Sarty and Rohwer, "Intelligence and the Air Forces," for use of Ultra and air power against U-boats in the last two years of the war.

28. Naval Historical Center, Operational Archives, Tenth Fleet, Anti-Submarine Measures (henceforth NHC/OA, Tenth Flt, ASM), Box 6. "Memorandum on Employment of HF/DF Equipment in Ships Employed on Anti-Submarine Operations," dictated at Boston Navy Yard, 11 August 1942 (hereafter HF/DF memo, 11 August 1942).

29. Howeth, *Communications-Electronics*, 401.

30. Fisher, *Race on the Edge of Time*, 306–7, 309.

31. The most detailed account of these events is in Deloraine, *When Telecom and ITT Were Young* (hereafter *Telecom*). Until the 1950s International Telephone and Telegraph was known as IT&T, but for clarity the modern usage—ITT—will be observed.

32. Deloraine, *Telecom*, 90–91, 117–32.

33. MacIntyre, *Battle of the Atlantic*, 137, 140–42. See also Pugh, "Military Need," 30; and Gannon, *Operation Drumbeat*, 398. For King's view see Lundeberg, "Allied Cooperation," 347.

34. MacIntyre, *Battle of the Atlantic*, 142–43.

35. MacIntyre, *Battle of the Atlantic*, 142–43. See also Rohwer's afterword in Dönitz, *Memoirs*, 504.

36. Sobel, *ITT*, 16–17.

37. Sampson, *The Sovereign State of ITT*, 27. Sampson, a journalist, has leveled many exaggerated and poorly substantiated charges against ITT's activities in World War II that have been essentially refuted by Sobel in his balanced and well-researched book. However, Sobel confirms what he terms "several of the more embarrassing Nixon/Chile episodes" in which ITT was involved. See Sobel, ix.

38. Morison, vol. 10, *Atlantic Battle Won*, 22–24.

39. Deloraine, *Telecom*, 119–20.

40. Pugh, "Military Need," 34. By 1945 British industrial capacity had declined by about 15 percent.

41. Jones, *Most Secret War*, 18. See also Niestlé, "German Technical and Electronic Development," summary, 449–50.

Chapter 2   The Target: Dönitz's Deadly Gray Wolves

1. Rhys-Jones, "The German System," 141–42.

2. Hessler, *U-Boat War*, vol. 1: 87, 98; vol. 2: 3, 29.

3. Lundeberg, "Allied Cooperation," 354. See also Rössler, "U-Boat Development," 124–37; and Dönitz, *Memoirs*, 299–314.

4. Padfield, *Dönitz*, 136–37, 154–55. See also Cremer, *U-Boat Commander*, 20.

5. Keegan, *Mask of Command*, 316–17.

6. Topp, "Manning and Training," 216–17.

7. Hackmann, *Seek and Strike*, 320–21. See also Rhys-Jones, "The German System," 140–45.

8. Padfield, *Dönitz*, 184.

9. Dönitz, *Memoirs*, 64.

10. Rhys-Jones, "The German System," 141–42.

11. Rahn, "The Campaign," 540. When convoys "disappeared" in 1941, "the Naval Intelligence Division was asked to investigate, but U-Boat Command took no immediate countermeasures."

12. Tidman, *The Operations Evaluation Group*, 334 n. 26 (hereafter *The OEG*).

13. McCue, *U-Boats in the Bay*, 170. See also Zimmerman, "Technology and Tactics," 488.

14. van der Vat, *Atlantic Campaign*, 281–82, 311–12, 319. See also Rohwer, *Critical Convoy Battles*, 103.

15. Dönitz, *Memoirs*, 22, 40–41.

16. Friedman, *Sub D & D*, 20, 42. See also Hackmann, *Seek and Strike*, 240.

17. Friedman, *Sub D & D*, 27. See also Hessler, *U-Boat War*, vol. 1: 109–19; vol. 2, Appendix: "Types and Specifications of U-Boats."

18. Degener-Böning, letter to author, 10 February 1991. Degener-Böning joined the Navy in 1938 at age 19.

19. Friedman, *Sub D & D*, 43. See also Hessler, *U-Boat War*, vol. 1: 109–19; vol. 2, Appendix.

20. Dönitz, *Memoirs*, 41.

21. Friedman, *Sub D & D*, 42–43. See also Dönitz, *Memoirs*, 108–9.

22. Dönitz, *Memoirs*, 115–16.

23. Padfield, *Dönitz*, 152–89. See also Dönitz, *Memoirs*, 18–24.

24. Dönitz, *Memoirs*, 107. See also Hinsley, *British Intelligence*, abridged edition, 128. Dönitz transmitted "comprehensive situation reports and patrol instructions." See also Rohwer, "The Wireless War," 408–17.

25. Dönitz, *Memoirs*, 30–31.

26. Rohwer, "The Wireless War," 408–17. See also Rohwer's afterword in Dönitz, *Memoirs*, 500, 502–4.

27. Dönitz, *Memoirs*, 63.

28. Hessler, *U-Boat War*, vol. 1: 21.

29. Kahn, *Seizing the Enigma*, 112–14.

30. Dönitz, *Memoirs*, 62. See also Rohwer and Hummelchen, *Chronology*, 23. In June 1940, seven U-boats of group Prien were deployed against convoy HX 48, which had been located through its WT (wireless telegraphy—radio) by B-Dienst, the cryptanalysis section of German naval intelligence.

31. Kahn, *Seizing the Enigma*, 112. In June 1940 "no regular, or even frequent solution of German naval cryptogram" was yet possible.

32. Kahn, *Seizing the Enigma*, 211–12. See also Hessler, *U-Boat War*, vol. 1: 57. "There, in the open Atlantic, the main problem was to locate the convoys."

33. Sternhell and Thorndike, *Antisubmarine Warfare*. Operations Evaluation Group Report No. 51 (hereafter OEG), 81.

34. Dönitz, *Memoirs*, 7. See also Hackmann, *Seek and Strike*, 234; and Rhys-Jones, "The German System," 145–46.

35. Dönitz, *Memoirs*, 134–35. See also Hessler, *U-Boat War*, vol 1: 62–64, 68; and Erskine, "U-Boats," 326.

36. Gannon, *Operation Drumbeat*, 80. See also Beesly, *Very Special Intelligence*, 54.

37. NHC/OA, Tenth Flt, ASM, Box 6. HF/DF memo, 11 August 1942. See also Erskine, "U-Boats," 324–28.

38. NHC/OA, Tenth Flt, ASM, Box 6. HF/DF memo, 11 August 1942.

39. Ibid. See also Rohwer, *Critical Convoy Battles*, 195.

40. Cremer, *U-Boat Commander*, 119.

41. Dönitz, *Memoirs*, 64.

42. Ibid.

43. Erskine, "U-Boats," 328 n. 7.

44. Perhaps the most detailed published account of U-boat radio procedures is in Wolfgang Hirschfeld's book *Feindfahrten*, which is based on his secret diary or log book. Hirschfeld served as radio operator in U-109 and then U-234 for the entire length of the war. On p. 356, for example, he records a lengthy and detailed late war (March 1944) message from BdU, which he calls "typical Dönitz."

45. Friedman, *Sub D & D*, 47. See also Holmes, *Double-Edged Secrets*, 72 n. 9.

46. Dönitz, *Memoirs*, 143.

47. Degener-Böning, letter of 10 February 1991. For a similar account of these procedures, see Kahn, *Seizing the Enigma*, 11; and Rohwer, *Critical Convoy Battles*, 34.

48. Degener-Böning, letter to author, 10 February 1991.

49. Gannon, *Operation Drumbeat*, 117–19. See also Hirschfeld, *Feindfahrten*, 259–60; and letters to author, 25 May and 14 June 1995.

50. NHC/OA, Tenth Flt, ASM, Box 6. HF/DF Course, 28 February 1943, 14–15. See also Erskine, "U-Boats," 329 n. 15, 324–25.

51. Hessler, *U-Boat War*, vol. 1: 77–78.

52. Werner, *Iron Coffins*, 265.

53. Guske, *War Diaries of U-764*, 173–74. For further information on frequency changes see Sternhell and Thorndike, OEG, 58; and Naval Research Laboratory (hereafter NRL) Report No. R-2229, of 5 February 1944, p. 3.

54. Rohwer and Hummelchen, *Chronology*, 194. See also Rohwer, "Die Auswirkungen der deutschen und britischen Funkaufklärung," 167–93.

55. Sadkovich, *Italian Navy*, 126–27.

56. Ibid.

57. Cremer, *U-Boat Commander*, 90.

58. David K. Brown, "Atlantic Escorts," 462.

59. NHC/OA, Tenth Flt, ASM, Box 6. HF/DF memo, 11 August 1942.

60. Erskine, "U-Boats," 324–28.

61. Cremer, *U-Boat Commander*, 63.

62. Hinsley, *British Intelligence,* 52–54.

63. Syrett, "Communications Intelligence," 49.

64. Ibid.

65. *Führer Conferences,* 230.

66. Sternhell and Thorndike, OEG, 19.

67. Hinsley, *British Intelligence,* 58.

68. van der Vat, *Atlantic Campaign,* 197.

69. Syrett, "Communications Intelligence," 58–59.

70. *Führer Conferences,* 232.

71. Ibid.

72. Capt. Benjamin T. Brooks, letter to author, 30 September 1993.

73. *Führer Conferences,* 239.

74. Captain Brooks, letter of 30 September 1993.

75. Syrett, "Communications Intelligence," 50.

76. Rohwer's afterword in Dönitz, *Memoirs,* 504.

77. Gannon's account of this U-boat campaign, *Operation Drumbeat,* puts most of the blame on Adm. Ernest King for failing to institute convoys and assign escorts to cover this route. For another view see Gardner, "An Allied Perspective," 518–19; and Love, "The U.S. Navy and Operation Roll of Drums."

78. Dönitz, *Memoirs,* 219.

79. Hinsley, *British Intelligence,* 379–80.

80. Ibid.

81. Syrett, *Defeat of German U-Boats,* 14.

82. Sternhell and Thorndike, OEG, 35, 81.

83. NHC/OA, Tenth Flt, ASM, Box 43. "A Study of the U-Boat Campaign," 21 July 1942, 4.

84. Beesly, *Very Special Intelligence,* 146. "It was ironical that, just as the Germans could not provide their U-boats with comprehensive orders before they sailed, and were compelled to issue frequent fresh instructions once they were at sea, so the Allies played into the hands of the B-Dienst in exactly the same manner."

85. Rohwer's afterword in Dönitz, *Memoirs,* 505.

86. Ibid.

87. Cremer, *U-Boat Commander,* 87–90.

88. Ibid., for example 90, 140, 167, 213.

89. Rahn, "The Campaign," 542.

90. Niestlé, "German Technical and Electronic Development," 438–39. See also Zimmerman, "Technology and Tactics," especially his summary on 488.

91. Rohwer, *Critical Convoy Battles,* 199.

92. Ibid.

93. Padfield, *Dönitz*, 230.

94. Salewski, "The Submarine War," in Lothar-Gunther Buchheim, *U-Boat War*, no p. nos.

95. Captain Brooks, letter of 30 September 1993.

96. Ibid.

97. Dönitz, *Memoirs*, 43–45.

98. Sternhell and Thorndike, OEG, 81.

99. Sternhell and Thorndike, OEG, 82. See also McKay, *Undersea Terror*, 123.

100. Rohwer and Hummelchen, *Chronology*, 191.

101. NHC/OA, Double Zero Files, Box 56. "Measures for Combatting the Submarine Menace," report by Combined Staff Planners, 1 March 1943.

102. Rohwer and Hummelchen, *Chronology*, 191.

103. Howse, *Radar at Sea*, 147–52, 180–81.

104. Rohwer's afterword in Dönitz, *Memoirs*, 506. See also Rohwer and Hummelchen, *Chronology*, 201.

105. Captain Brooks, letter of 30 September 1993.

106. NHC/OA, Double Zero Files, 1942–47, Box 21. Pencil memo, "Naval Strategy in the Atlantic and Pacific," by Fleet Adm. Ernest J. King, USN, 18 August 1943.

107. Syrett, *Defeat of German U-Boats*, 22, explains the invaluable role of Ultra decrypts.

108. NHC/OA, Research Section, Post-1974 Command File. On the need to cover the sources of Ultra, see R. F. Cross Associates, Ltd., *Sea-Based Airborne Antisubmarine Warfare, 1940–1977* (hereafter Cross report), vol. 1: ix. For analysis of the problem of reconstructing how exactly each piece of radio intelligence was used operationally, and to what effect, see Syrett, "Communications Intelligence."

109. See Middlebrook, *Convoy*, 311n, for the *Bogue's* number of kills. See Table 9-1 for HF/DF-assisted attacks.

110. Sternhell and Thorndike, OEG, 82. See also Gretton, *Crisis Convoy*, 148–53.

111. Cremer, *U-Boat Commander*, 139.

112. NHC/OA, King Papers, Box 13. Miscellaneous Materials, 1943–48, "Resume of Anti-Submarine Operations Against the German U-Boats in World War II," F. S. Low to King (hereafter Low Resume), 28 October 1944.

113. Cross report, vol. 3: 160.

114. Cremer, *U-Boat Commander*, 138–39, 145. "Of ten U-tankers only two survived that summer."

115. Rohwer and Hummelchen, *Chronology,* 215.

116. Hinsley, *British Intelligence,* 383–84.

117. Till, "Battle of the Atlantic," 589.

118. Sternhell and Thorndike, OEG, 82.

119. Dönitz, *Memoirs,* 420–23.

120. Hinsley, *British Intelligence,* 384–85, 597–98.

121. Niestlé, "German Technical and Electronic Developments," 438.

122. Deloraine, *Telecom,* 123.

123. Professor Rohwer, letter to author, 9 May 1994. Professor Rohwer writes that "the Admiralty had in 1944 great fears for a general introduction of this Kurier." See also Rohwer, "The Wireless War," 415–16.

124. Hinsley, *British Intelligence,* 597.

125. Lundeberg, "Allied Cooperation," 364.

126. Hinsley, *British Intelligence,* 390. In fairness to Dönitz it should be pointed out that no other country succeeded in producing a true submarine in the course of the war.

127. Rohwer and Hummelchen, *Chronology,* 353.

128. Bond, *War and Society,* 178.

129. Barnett, *Engage the Enemy,* 575.

Chapter 3   Sonar, Radar, and Ultra

1. NHC/OA, King Papers, Box 13. F. S. Low Memorandum, Enclosure A, "Appreciation of the Anti-Submarine Situation, 20 April 1943" (hereafter Low Memorandum).

2. Hezlet, *Electronics,* 43–49.

3. Hessler, *U-Boat War,* vol. 3: 34–35. See also Dönitz, *Memoirs,* 339–41, 408.

4. Hessler, *U-Boat War,* vol. 3: 51.

5. I am indebted to Prof. Jürgen Rohwer (letter to author, 30 December 1994) for information on Dönitz's use of Hessler's study in the writing of his memoirs in the 1950s. At that time, Dönitz believed that "the supremacy which the enemy's defences had enjoyed over the U-boat since 1943 . . . had been due to the introduction of the ultra shortwave surface radar." *Memoirs,* 427.

6. I am grateful to Bernard F. Cavalcante, branch head, Operational Archives, Naval Historical Center, for information on postwar archivists.

7. Cross report, vol. 2: 18–19.

8. Dow, "Navy Radio and Electronics," 284. During the war Commodore Dow was the Director of the Electronics Division, Bureau of Ships.

9. Howeth, *Communications-Electronics,* 518–19.

10. Hackmann, *Seek and Strike,* 257–58; Howse, *Radar at Sea,* 2.

11. Captain Brooks, letter to Captain Scott Lothrop, 7 March 1992.

12. Elliott, *Allied Escort Ships*, 327–28. See also Dow, "Navy Radio and Electronics," 285; and U.S.Navy, Bureau of Ships, Naval Research Laboratory, Anacostia, Washington, D.C., unpublished World War II Administrative History No. 134, "War History of the Naval Research Laboratory," 1 November 1946 (hereafter NRL war history), 9.

13. Love, *History of the U.S. Navy*, vol. 2: 74.

14. NRL war history, 9. See also Dow, "Navy Radio and Electronics," 285.

15. Barnett, *Engage the Enemy*, 44–45. See also Hezlet, *Electronics*, 179.

16. Capt. Benjamin T. Brooks, USN (Ret.), interview with author, 9 April 1992.

17. Ibid.

18. Ibid.

19. NHC/OA, Action Report, Box 856, No. 0045. Task Group (TG) 22.3, 11 October 1944.

20. Brooks interview, 9 April 1992.

21. Hessler, *U-Boat War*, vol. 2: 47; vol. 3: 78.

22. Meigs, *Slide Rules*, 145–46.

23. For naval radar see Howse, *Radar at Sea*.

24. van der Vat, *Atlantic Campaign*, 118. See also Page, "Radar Model XAF," 88.

25. Page, "Radar Model XAF," 88.

26. Hezlet, *Electronics*, 171–72. See also Watson-Watt, *Three Steps*, 123–25.

27. Watson-Watt, *Three Steps*, 124.

28. Ibid., 125.

29. Fisher, *Race on the Edge of Time*, 29–31. Watson-Watt is often accorded all the credit for radar. Fisher sets the record straight on the cooperative efforts that went into its early development. See also Howse, *Radar at Sea, 2*.

30. NRL war history, 9. The definitive work on the development of radar at the NRL is Allison, *New Eye for the Navy*.

31. NRL war history, 43. See also Allison, *New Eye for the Navy*, 97; Howeth, *Communications-Electronics*, introduction by Adm. Nimitz, xiv.

32. Van Atta and Silver, "Contributions to the Antenna Field," 692–97.

33. Ibid.

34. Howse, *Radar at Sea*, 79–80, 86–87.

35. Ibid. See also Allison, *New Eye for the Navy*, app. G.

36. Quoted in Hezlet, *Electronics*, 114.

37. Abbazia, *Mr. Roosevelt's Navy*, 276, notes that escorts with convoy SC 48 were "blind" without any radar. See also Rohwer, *Axis Submarine Successes*, 69; and Rohwer and Hummelchen, *Chronology*, 90–91.

38. Hezlet, *Electronics*, 212.

39. Gebhard, *Naval Radio-Electronics,* 170.

40. Page, "Radar Model XAF," 88. See also Hezlet, *Electronics,* 217–19; and Howse, *Radar at Sea,* 169.

41. Hezlet, *Electronics,* 211, 217–19. See also Brooks letter of 30 September 1993.

42. NHC/OA, Action Report, Box 856, No. 0045. TG 22.3, 11 October 1944. It was brought out at a conference of TG 22.3 in August that seagulls "gave good radar 'pips.'" Also the comments appended to the action report noted that while air and surface craft radar functioned well, "the a/c had numerous disappearing 'blips.'" See also NHC/OA, Action Report, Box 171, No. 083. TG 41.7, 11 September 1944. Disappearing "contacts" seem to have been caused by clouds or by other aircraft, it is noted.

43. NHC/OA, Action Report, Box 1240, No. 0026. TG 22.1, 25 November 1944.

44. Hezlet, *Electronics,* 113.

45. Howse, *Radar at Sea,* 45.

46. Hessler, *U-Boat War,* vol. 1: 81–82.

47. Rohwer and Douglas, "Canada and the Wolf Packs," 160.

48. Hessler, *U-Boat War,* vol. 2: 26, 42, covers Metox. Vol. 3: 4, covers Hagenuk and Naxos. For the ban on the use of Metox, see vol. 3: 21. For further details on Naxos and Tunis, see vol. 3: 57.

49. Fisher, *Race on the Edge of Time,* 271–72.

50. NHC/OA, Tenth Flt, ASM, Box 12. SECNAV to ALNAV, 2 August 1943.

51. NHC/OA, Action Report, Box 896 (no serial no.). TG 21.14, 10 September 1943.

52. Howse, *Radar at Sea,* 169, 201. Ruler-class U.S.-built ships had U.S. radar, while *Archer* class had British radar.

53. See Howse, *Radar at Sea,* 169, 251, for a comparison of British and American sets.

54. Rohwer and Hummelchen, *Chronology,* 199–200.

55. Ibid., 211–12.

56. Quoted in MacIntyre, *Battle of the Atlantic,* 194.

57. Ibid.

58. McCue, *U-Boats in the Bay,* 172.

59. Kahn, *Seizing the Enigma,* 36–37, 52, 62, 65, 86–87.

60. Ibid., 112, 114.

61. Beesly, *Very Special Intelligence,* 68–69. See also van der Vat, *Atlantic Campaign,* 73–74; and McCue, *U-Boats in the Bay,* 33.

62. Rohwer, "The Wireless War," 408–17.

Chapter 4   The Secret Weapon: Huff Duff

1. Hezlet, *Electronics*, 163. See also Howeth, *Communications-Electronics*, 527; and Barnett, *Engage the Enemy*, 8.

2. Keen, *Wireless Direction Finding*, 466. Keen is still the standard text on direction finding.

3. U.S. Patent Office, Patent No. 984,108, 14 February 1911. See also Kahn, *Seizing the Enigma*, 144–45; and Hezlet, *Electronics*, 214.

4. Keen, *Wireless Direction Finding*, 468.

5. Gething, "High-Frequency Direction-Finding," 49.

6. Keen, *Wireless Direction Finding*, 590–91, 610–11.

7. Howeth, *Communication-Electronics*, 532, 601–2.

8. Keen, *Wireless Direction Finding*, 10–11.

9. Adams and Colin, "Frequency, Power and Modulation," 144. See also Hezlet, *Electronics*, 161.

10. Keen, *Wireless Direction Finding*, 10–11.

11. Gething, "High-Frequency Direction-Finding," 49. See also Norton, "Radio-Wave Propagation," 698–704.

12. Watson-Watt, *Three Steps*, 60–61, 359.

13. Watson-Watt and Herd, "An Instantaneous Direct-Reading Radiogoniometer," 611. See also Busignies Papers.

14. H. S. Black and J. O. Edson of Bell Telephone Laboratories Incorporated, patent No. 2,083,495, 8 June 1937; Henri Gaston Busignies, patent No. 2,208,209, 16 July 1940; Donald S. Bond, patent No. 2,234,331, 11 March 1941; Josef Plebanski, patent No. 2,284,475, 26 May 1942.

15. Gething, "High-Frequency Direction-Finding," 49–50.

16. Van Atta and Silver, "Contributions to the Antenna Field," 692.

17. Keen, *Wireless Direction Finding*, 520–29.

18. De Walden et al., "High-Frequency Cathode-Ray Direction Finder," 824, 827, 830–34.

19. Ibid.

20. Howeth, *Communications-Electronics*, 602. See also Howse, *Radar at Sea*, 143.

21. Erskine, "U-Boats," 324–30.

22. Hezlet, *Electronics*, 189. See also Rohwer, *Critical Convoy Battles*, 16–22; and Syrett, "Communications Intelligence," 47.

23. Beesly, *Very Special Intelligence*, 10–14, 16–17. See also Kahn, *Seizing the Enigma*, 144–45. Direction-finding posts were eventually established in the Shetlands, Wick, and Cupar in Scotland; and in Scarborough, Winchester, Chelmsford, Lydd, and Land's End in England.

24. Gething, "High-Frequency Direction-Finding," 49–50. See also Keen, *Wireless Direction Finding*, 406–7.

25. Gething, "High-Frequency Direction-Finding," 49–50.

26. Hezlet, *Electronics*, 177, 189.

27. Rohwer, *Critical Convoy Battles*, 229.

28. Gannon, *Operation Drumbeat*, 167. See also Abbazia, *Mr. Roosevelt's Navy*, 297, 353.

29. Hackmann, *Seek and Strike*, 259, 265. See also Fisher, *Race on the Edge of Time*, 267–69.

30. Watson-Watt, *Three Steps*, 335.

31. Milner, *North Atlantic Run*, 112–13. See also Watson-Watt, *Three Steps*, 205, 363; and Howse, *Radar at Sea*, 224–25, 305. For a detailed account of Canadian naval technology, see Zimmerman, *Great Naval Battle*.

32. Syrett, "Communications Intelligence," 47.

33. Erskine, "U-Boats," 325.

34. Dr. Maxwell K. Goldstein, undated memoir, "Some Research and Development Experiences in the U.S. Navy," hereafter Goldstein memoir.

35. Keen, *Wireless Direction Finding*, 407, 470–71.

36. Van Atta and Silver, "Contributions to the Antenna Field," 692.

37. NHC/OA, Tenth Flt, ASM, Box 6. HF/DF memo, 11 August 1942.

38. Howse, *Radar at Sea*, 144.

39. Keen, *Wireless Direction Finding*, 510. See also Watson-Watt, *Three Steps*, 361.

40. NHC/OA, Tenth Flt, ASM, Box 6. HF/DF memo, 11 August 1942.

41. Watson-Watt, *Three Steps*, 361–62.

42. Ibid. See also Howse, *Radar at Sea*, 146.

43. NRL Report No. R-1896, 20 June 1942, 1.

44. Goldstein memoir.

45. Keen, *Wireless Direction Finding*, 473, 478–79, 510. See also Hague, *The Towns*, 20; and De Walden et al., "High-Frequency Cathode-Ray Direction Finder," 824–26.

46. Goldstein memoir.

47. Gething, "High-Frequency Direction-Finding," 49–50. See also De Walden et al., "High-Frequency Cathode-Ray Direction Finder," 823, 836.

48. Morison, vol. 1, *Battle of the Atlantic*, 105 n. 38.

49. Gebhard, *Naval Radio-Electronics*, 299.

50. Howse, *Radar at Sea*, 146.

51. NHC/OA, Tenth Flt, ASM, Box 6. HF/DF memo, 11 August 1942. See also Raven and Roberts, *British Battleships*, 390.

52. Howse, *Radar at Sea*, 144.

53. NHC/OA, Tenth Flt, ASM, Box 6. Comdr. TF 24 to CINCLANT, 27 October 1942.

54. Milner, *North Atlantic Run,* chap. 1. See also Milner, *U-Boat Hunters,* 269–79; Douglas and Rohwer, "Most Thankless Task," 210–21; and Rohwer and Douglas, "Canada and the Wolf Packs," 181.

55. Douglas and Rohwer, "Most Thankless Task," 210–21. See also Rohwer and Hummelchen, *Chronology,* 170–71.

56. Hezlet, *Electronics,* 230.

57. Howeth, *Communications-Electronics,* 400–401, quotes Admiral King, 8 December 1945.

58. NHC/OA, Tenth Flt, ASM, Box 6. HF/DF memo, 11 August 1942.

59. Howse, *Radar at Sea,* 179, 277.

60. NHC/OA, Double Zero Files, Box 56. "Measures for Combatting the Submarine Menace," report by Combined Staff Planners, 1 March 1943, and Appendix B.

61. Gretton, *Crisis Convoy,* 168.

62. Friedman, *Sub D & D,* 48. See also NHC/OA, Tenth Flt, ASM, Box 6. HF/DF memo, 11 August 1942, states that "HF/DF offers the only means of establishing the presence of a U-boat before it is in a position to attack a convoy."

Chapter 5   The Inventor: Henri Busignies

1. Deloraine, "L'Histoire du Huff-Duff" (hereafter "Huff-Duff"). See also author's 23 March 1993 interview with Ruth Lockhart Lombardi, Henri Busignies's longtime assistant.

2. Deloraine, "Huff-Duff," 1. See also interview with Ruth Lombardi, 23 March 1993. Henri Busignies died in 1981, while on vacation in the south of France. At that time he still had an office permanently at his disposal at the ITT plant in Nutley, NJ.

3. Unsigned typescript memoir of Busignies's career in electronics (hereafter electronics memoir), Busignies Papers.

4. Electronics memoir, 1.

5. "Contributors to This Issue," *Electrical Communication* (hereafter *EC*), 23 (March 1946): 109. *EC* is "the technical journal of the International Telephone and Telegraph Corporation and associate companies."

6. Interview with Ruth Lombardi, 21 August 1991.

7. Busignies, electronics memoir, 1–2.

8. Interview with Ruth Lombardi, 13 November 1992.

9. Mme Busignies letter to author, 19 June 1993. See also electronics memoir, 9.

10. Richardson, "Galena," 7.

11. Ibid. See also Busignies, "Communication and Navigation," 592. In 1961 Busignies predicted the use of laser beams, satellite television transmission, electronic mail and newspapers, and much more.

12. Watson-Watt's paper, originally published in the *Journal of the Institute of Electrical and Electronic Engineers,* is cited in De Walden, "Development of a High-Frequency Cathode-Ray Direction Finder for Naval Use," 837. See Sobel, 106, for description of Busignies.

13. Busignies, electronics memoir, 1–2.

14. Busignies, speech to the Radio Club of America, 1975.

15. "Contributors to this Issue," *EC* 23 (March 1946): 109

16. "Huff-Duff Presented for Smithsonian Display," *WITTS* (Within the International Telephone and Telegraph System), February 1976: 1. See also Busignies, "The Automatic Radio Compass," *EC,* 157–72.

17. Auphan and Mordal, *La Marine Française,* 35–37. In 1940 Capt. (later Rear Admiral) Paul Auphan served as deputy chief of staff in Adm. Jean Darlan's wartime headquarters.

18. "Electrical Communication: 1940–1945, War Years' Review," in 3 parts. *Electrical Communication,* 1946 (henceforth "War Years' Review," *EC*).

19. Deloraine, "Huff-Duff," 2. See also NRL Report No. R-2229, 5 February 1944, "Interpolation Charts for HF-DF Shipboard Calibration."

20. "War Years' Review," part 2. *EC,* 23 (June 1946): 214–16.

21. Richardson, "Galena," 7.

22. Deloraine, *Telecom,* 1–37.

23. Ibid.

24. Ibid., 98–99. See also Sobel, *ITT,* 31–32.

25. Deloraine, *Telecom,* v–vi.

26. Busignies, electronics memoir, 2.

27. Deloraine, *Telecom,* 107–8.

28. Auphan and Mordal, *La Marine Française,* 42–51.

29. Watson-Watt, *Three Steps,* 205–6. Hezlet, *Electronics,* 171, confirms the development of the French high-power pulse system with British help.

30. Auphan and Mordal, *La Marine Française,* 42–51.

31. Deloraine, *Telecom,* 107–8.

32. Watson-Watt, *Three Steps,* 207.

33. Auphan and Mordal, *La Marine Française,* 61.

34. Watson-Watt, *Three Steps,* 208.

35. Deloraine, *Telecom,* 108.

36. Ibid.

37. Watson-Watt, *Three Steps,* 208. When Watson-Watt told himself that this time the French would not hold, he was referring to General Pétain's fa-

mous boast at Verdun in 1916: *"Ils ne passeront pas"* (they [the Germans] will not pass).

38. Watson-Watt, *Three Steps*, 338–40.

39. Deloraine, *Telecom*, 108. See also Hezlet, *Electronics*, 171, for French radar on the *Normandie*. Unlike Deloraine, Hezlet describes it as very primitive. Allison, *New Eye for the Navy*, 93, refers to "a crude form of continuous-wave radar" on the *Normandie*.

40. Deloraine, *Telecom*, 108–9. See also Busignies Papers, biographical sketch, typed manuscript, April 1973.

41. Interview with Mme Busignies, 12 June 1991.

42. Ibid.

43. Deloraine, *Telecom*, 109–10. See also interview with Mme Busignies, 12 June 1991.

44. Deloraine, *Telecom*, 110–12. See Walton, *Miracle of World War II*, 445, for the Nazi seizure of European plants owned by Americans (including ITT) long before the United States was in the war.

45. Deloraine, *Telecom*, 111–12.

46. Interview with Mme Busignies, 12 June 1991.

47. Deloraine, *Telecom*, 112.

48. Deloraine, *Telecom*, 112–13. See also Busignies, electronics memoir, 3; and interview with Mme Busignies, 12 June 1991.

49. Deloraine, *Telecom*, 113–15.

50. Ibid.

51. Interview with Mme Busignies, 12 June 1991.

52. Ibid.

53. Ibid.

54. Ibid.

55. Deloraine, "Huff-Duff," 3. Sumner Welles was at that time acting secretary of state.

56. Interview with Mme Busignies, 12 June 1991.

57. Ibid.

58. Ibid. See also Deloraine, "Huff-Duff," 3. St. Exupéry was the author of the well-known children's book *Le Petit Prince*, among other works. Hyde, *Room 3603*, 95–97, is an amusing account of other searches of the luggage of Vichy diplomats returning to Europe, searches that were conducted by the British in Bermuda, officially or otherwise.

59. Deloraine, *Telecom*, 115.

60. Ibid. See also Busignies, electronics memoir, 4; and Rear Adm. Frederick R. Furth, USN (Ret.), letter of 30 May 1992. Both Admiral Furth and General Colton became vice presidents of ITT on retirement.

61. Deloraine, *Telecom*, 116–17.

62. Ibid. "Recent Telecommunications Developments," *EC*, 241–43, explains that every land station needed four HF/DF devices, each tuned to a different part of the spectrum.

63. Sivowitch, letter to author, 12 August 1991.

64. Deloraine, *Telecom*, 117. See also Deloraine, "Huff-Duff," 4–5.

65. Goldstein memoir. See also "Vita of Dr. M.K. Goldstein, Naval Research Laboratory" (hereafter Goldstein vita).

66. Gannon, *Operation Drumbeat*, 164, maintains that two U.S. naval officers returned from a technical mission to Britain in April 1941 with a very advanced HF/DF device. If accurate, this might have impacted on the hiatus with regard to orders for Busignies's D/F. The two officers, a lieutenant and an ensign, may not have had the technical background to judge the equipment properly.

67. NRL Report No. R-1896, 20 June 1942, 15.

68. Deloraine, *Telecom*, 118–19.

69. Sobel, *ITT*, 34–35.

70. Sampson, *Sovereign State of ITT*, 26–27.

71. Interview with Mme Busignies, 12 June 1991.

72. Deloraine, *Telecom*, 119.

73. Ibid.

74. Quoted in Hyde, *Room 3603*, 100–101.

75. Hyde, *Room 3603*, 1–3.

76. Interview with Mme Busignies, 12 June 1991.

77. National Archives and Records Administration, Suitland, MD, Record Group 19, Bureau of Ships, General Correspondence (hereafter NARA, RG 19, BuShips, Gen. Corr.), 1940–45, Box 617.

78. NARA, RG 19, BuShips, Gen. Corr., Box 617. Behn to Noyes, 14 August 1941.

79. Ibid.

80. Ibid., Buttner to Colton, 14 August 1941.

81. Ibid., Knox to Behn, 2 May 1943.

82. Ibid.

83. Ibid., Behn to Knox, 6 May 1943.

84. Ibid.

85. Ibid., Knox to Behn, 10 May 1943.

86. Ibid., director of naval intelligence to director of base maintenance division, 13 January 1943.

87. Ibid., case history, 1.

88. Ibid., 2–3.

89. Ibid.

90. Sampson, *Sovereign State of ITT*, 28–40. See also Martin, *All Honorable*

Men, 209. ITT owned a group of corporations (including Lorenz) that made up the third largest electrical combine in Germany, after Siemens and A.E.G.

91. Sobel, *ITT,* ix.

92. Ibid., 106–9.

93. Ibid., 111–12.

94. Deloraine, *Telecom,* 40.

95. Sobel, *ITT,* 18, 85. See also Sampson, *Sovereign State of ITT,* 28.

96. NARA, RG 19, BuShips, Gen.Corr., Box 617. Case history, 4–6.

97. Ibid.

98. Ibid.

99. Ibid.

100. Ibid., 7.

101. NRL war history, 167–68.

102. NARA, RG 19, BuShips, Gen. Corr., Box 617. International Telephone and Radio Laboratory to BuShips, 22 January 1942.

103. NHC/OA, Ready Reference Files. Eller Letters, Eller to Coates, 31 July 1963.

104. Deloraine, "Huff-Duff," 5. See also Richardson, "Galena," 6–7.

105. Deloraine, "Huff-Duff," 5–6.

106. Ibid.

107. NHC/OA, Ready Reference Files. Eller Letters, Eller to Coates, 31 July 1963.

108. Deloraine, *Telecom,* 119–20.

109. Ibid. See also Goldstein memoir.

110. NRL Report No. R-1896, 20 June 1942, "The Performance of the British FH3 and the International Telephone and Radio Laboratories [*sic*] Crossed Loop Type High Frequency Direction Finders (Aboard USS *Corry* DD 463)," by Maxwell Goldstein. See Abstract, i. The photographs are clearly labeled as if showing both the British and the American devices. Yet, according to several of the ITT engineers who worked on Busignies's DAQ model HF/DF, the only D/F loop to appear in the photos, other than the British FH3, is a (round) rotating medium frequency-type antenna.

111. NHC/OA, Tenth Flt, ASM, Box 13. VCNO to Chief, BuShips, 26 June 1943. "Type Installations for Ships, Part VI, General Characteristics of Direction Finders."

112. Goldstein memoir.

113. "War Years' Review," part 2. *EC,* 23 (June 1946): 214.

114. Interview with Avery Richardson, 19 February 1991.

115. "War Years' Review," part 2. *EC,* 23 (June 1946): 214.

116. Dow, "Navy Radio and Electronics," 284. See also NRL war history, 201–18; and NHC/OA, Tenth Flt, ASM, Box 6. HF/DF Course Manual, 28 February 1943, 11.

117. Deloraine, "Telephone, Telegraph, and Telex Traffic," *EC*, 39 (1964): 276. See also Busignies Papers, biographical sketch, 1973.

118. Deloraine, "Telephone, Telegraph, and Telex Traffic," *EC*, 39 (1964): 276. See also Busignies Papers, biographical sketch, 1973.

Chapter 6   The Decision Makers: Individuals and Organizations

1. Rear Adm. A. H. Van Keuren, "The U.S. Naval Research Laboratory" (hereafter "The NRL"), 221.

2. For Roosevelt's role in the Atlantic campaign before December 1941, see Abbazia, *Mr. Roosevelt's Navy*.

3. Farago, *Tenth Fleet*, 42–44.

4. Y'Blood, *Hunter-Killer*, 9. See also O'Neill, *Democracy at War*, 150–51.

5. Meigs, *Slide Rules*, 64–65.

6. van der Vat, *Atlantic Campaign*, 233–34, 239.

7. Not everyone agrees with this picture. Love, "The U.S. Navy and Operation Roll of Drums," 120, calls King's Anglophobia "unproven, yet widely alleged."

8. Among the most vocal critics, see Gannon, *Operation Drumbeat*, 398.

9. Hoopes and Brinkley, *Driven Patriot*, 171.

10. Meigs, *Slide Rules*, 89.

11. O'Neill, *Democracy at War*, 143, 146–48.

12. Hoopes and Brinkley, *Driven Patriot*, 163.

13. NARA, RG 19, BuShips, Gen. Corr., Box 1061. CINCLANT to CNO, 22 February 1942.

14. Ibid.

15. Howeth, *Communications-Electronics*, 435.

16. Hoopes and Brinkley, *Driven Patriot*, 171.

17. Allard, "A United States Overview," 569–70.

18. Ibid., 568–69. See also Lundeberg, "Allied Cooperation," 356–57.

19. NHC/OA, Double Zero Files, Box 56. "Measures for Combatting the Submarine Menace," report by Combined Staff Planners, 1 March 1943.

20. Meigs, *Slide Rules*, 213. See also Love, *History of the U.S. Navy*, vol. 2, 75. Love credits Captain Carney of the Atlantic Fleet Support Force with the idea for the Anti-Submarine Warfare Operational Research Group (ASWORG), stating that King gave the task to Baker.

21. Hackmann, *Seek and Strike*, 282, discusses the influence of Captain Baker on the development of a fast-sinking depth charge, June 1942.

22. For the most comprehensive account of the OIC, see Patrick Beesly, *Very Special Intelligence*.

23. Farago, *Tenth Fleet*, 141–42.

24. Sternhell and Thorndike, OEG, foreword, vii.

25. NHC/OA, King Papers, Box 13. Low Resume, 28 October 1944, Enclosure J, "Digest of Minutes of Conference on ASW held at Headquarters, COMINCH, 25 January 1944."

26. Gannon, *Operation Drumbeat*, 341, 384. There were many U.S. Navy officers making the pilgrimage to Britain at this time to see what they could learn. Among them was Comdr. Kenneth A. Knowles, who was later to head the Atlantic Section of the Combat Intelligence Division of Tenth Fleet.

27. Van Keuren, "The NRL," 221. The best work on the NRL, and especially its role in the development of radar, is Allison, *New Eye for the Navy*.

28. NRL war history, 4–5.

29. Dr. David van Keuren, letter to author, 15 March 1991. See also NRL war history, 265–67; Allison, *New Eye for the Navy*, 89–92.

30. NRL war history, 209.

31. Ibid., 222, quoting Admiral Van Keuren.

32. Ibid., 44–45, 276–77.

33. Ibid., 278.

34. Ibid., 31–32, 278.

35. Ibid., 278.

36. Ibid., 276–77.

37. Ibid., 278.

38. Rear Admiral Furth, letter to Capt. Scott Lothrop, 30 May 1992, and author's interview, 10 June 1992. Admiral Furth was adviser on electronics to Admiral Noyes, and he was a naval member of most of the wartime scientific committees, including the Combined Chiefs of Staff Radar Development Committee, the NRL Priorities Committee, ANEPA (Army-Navy Electronic Production Agency), and the NDRC, where he served with Admiral Furer of the Office of the Coordinator for Research and Development (OCRD).

39. NRL war history, 35.

40. Goldstein memoir.

41. Van Keuren, "The NRL," 224–25.

42. Ibid. See also NRL war history, 21, 57–58.

43. NRL war history, 134, 141.

44. Ibid., 15, 37, 55–57, 162.

45. Ibid., 45, 92, 209. See also "Institute News and Radio Notes," *IRE*, 34 (July 1946): 477.

46. Meigs, *Slide Rules*, 200. See also Hackmann, *Seek and Strike*, 251–52. He notes that submarine research was being conducted in some seventy establishments throughout the country.

47. Hackmann, *Seek and Strike*, 254–56. See also Meigs, *Slide Rules*, 123.

48. NRL war history, 92–94.

49. Ibid., 279.

50. Ibid. See also Van Keuren, "The NRL," 221–22.

51. NRL war history, 62.

52. Ibid., 36–37.

53. Ibid., 38, 44–47.

Chapter 7   The Expediters: Procurement and Supply

1. Walton, *Miracle of World War II*, 28–29, and 536 for Goebbels.

2. Richardson, "Galena," 2–8.

3. Ibid.

4. NARA, RG 19, BuShips, Gen. Corr., Box 1060. Rockett to Radio Division, Navy Dept., 3 July 1940.

5. Gardner, "An Allied Perspective," 519.

6. NARA, RG 19, BuShips, Gen. Corr., Box 1063. Circular Letter No. 26 from Chief, BuShips, 10 December 1942.

7. Howeth, *Communications-Electronics*, 402, 422.

8. Love, *History of the U.S. Navy*, vol. 2, 74.

9. NARA, RG 19, BuShips, Gen. Corr., Box 1061, 14 July 1941.

10. NARA, RG 74, Bureau of Ordnance (BuOrd), Box 99. Chief, BuShips, to CINCLANT, 10 November 1941.

11. NARA, RG 19, BuShips, Gen. Corr., Box 1061. Shipment Order from Commandant, Navy Yard, Charleston, S.C., to Supply Officer, Navy Yard, Washington, D.C., for Radio Model DF Direction Finder, 13 June 1941.

12. NARA, RG 19, BuShips, Gen. Corr., Box 1061. Radio Communication Service to BuAero, 20 December 1941.

13. NRL Report No. R-2229, 5 February 1944, 13. Converted British FH3s became American DARs.

14. NARA, RG 19, BuShips, Gen. Corr., Box 1062. Skrainka to Bureau of Radio Communication, 3 August 1942.

15. NARA, RG 19, BuShips, Gen. Corr., Box 1062. Asst. CNO to Asst. Chief of Staff, 20 August 1942.

16. Richardson, "Galena," 5–6.

17. O'Connell, Pachynski, and Howeth, "Summary of Military Communication," 1246.

18. Hackmann, *Seek and Strike*, 256, puts the number of OSRD and NDRC scientists at 32,000. See Howeth, *Communications-Electronics*, 542–43, for NDRC contracts.

19. NARA, RG 19, BuShips, Box 452. BuShips to Navy Rep. WPB Area Production Urgency Committee, New Brunswick–Paterson, N.J., 11 May 1945.

20. Howeth, *Communications-Electronics*, 421, 542–43.

21. NARA, RG 19, BuShips, Gen. Corr., Box 617. BuShips to FCC, 2 January 1942.

22. Interview with Avery Richardson, 19 February 1991.

23. Rear Admiral Furth, letter to author, 30 May 1992. In 1940, then-Lieutenant Commander Furth was assigned to OP-20, the fleet section of Naval Communications. Later, "as electronics continued to become more and more widespread and played a more important role in the war, an Electronics Division in the Office of VCNO was created as OP-25, and I was made the Deputy Director."

24. Connery, *The Navy and Industrial Mobilization*, 56–65.

25. Walton, *Miracle of World War II*, 484–85. "Parts of some of the greatest military secrets were spread through thousands of shops and factories, both large and small."

26. Avery Richardson, letter to author, 19 June 1991.

27. NARA, RG 19, BuShips, Gen. Corr., Box 1061. Memo from Office of Inspector of Naval Material, N.Y., 12 December 1941.

28. NARA, RG 19, BuShips, Gen. Corr., Box 1070, memo from Chief of Procurement and Material to Chief BuShips, 24 March 1943. See also Howeth, *Communications-Electronics*, 544–45.

29. NARA, RG 19, BuShips, Gen. Corr., Box 1062. Chief, Office of Procurement and Material, to Chiefs, BuShips, BuOrd, BuAero, Bureau of Yards and Docks, 8 August 1942.

30. Ibid.

31. Howeth, *Communications-Electronics*, 544–45.

32. Ibid., 422, 544–45.

33. NARA, RG 19, BuShips, Gen. Corr., Box 1062. Hooper to Chief, BuShips, 23 July 1942.

34. Ibid., Box 1063. Hooper to Chief, BuShips, 4 December 1942.

35. Ibid.

36. Ibid.

37. NARA, RG 19, BuShips, Gen. Corr., Box 452. Chief, BuShips, to Federal Telephone and Radio Laboratories (FT & RL), New York, N.Y., regarding certificates of deferment for two employees, 10 May 1945.

38. Interview with Eugene Lombardi, 3 August 1995. Lombardi returned to ITT at the end of the war.

39. Goldstein Papers, letter from Bowen to Goldstein concerning draft status, 23 May 1941.

40. Goldstein Papers, letter concerning draft status from Briscoe to Selective Service, Local Board No. 24, 21 November 1941.

41. NRL war history, 103–4. After the war Goldstein left the NRL to establish his own company, Balco Research Laboratories. He took with him twelve other NRL Ph.D.s.

42. NARA, RG 19, BuShips, Gen. Corr., Box 1062. Hooper to Chief, BuShips, 23 July 1942.

43. NRL Report No. R-1896, 20 June 1942, 10.

44. Interviews with Avery Richardson, 12 and 19 February 1991.

45. Ibid.

46. Goldstein memoir.

47. Goldstein Papers, Johns Hopkins Yearbook, 1930.

48. Goldstein vita; Goldstein resume; and Goldstein memoir.

49. Goldstein memoir.

50. Ibid.

51. Goldstein Papers. See also citation.

52. Howeth, *Communications-Electronics*, 401–2.

53. Ibid., 401–2, 436.

54. Ibid., 438.

55. Interview with Avery Richardson, 19 February 1991.

56. "War Years' Review," part 1. *EC*, 23 (March 1946): 14.

57. NRL war history, 88–89.

58. Interview with Avery Richardson, 19 February 1991.

59. NARA, RG 19, BuShips, Gen. Corr., Box 1062. Hooper to Chief, BuShips, 23 July 1942.

60. Ibid.

61. Ibid.

62. NHC/OA, King Papers, Box 13. Low Resume, 28 October 1944.

63. Ibid.

64. Avery Richardson, letter to author, 19 June 1991. Author's interview with Eugene Lombardi, 29 January 1991, confirms "there was good cooperation on the technical level" between ITT and the navy.

65. NARA, RG 19, BuShips, Gen. Corr., Box 1070. Office of Inspector of Naval Material, N.Y., to BuShips, 6 March 1943.

66. Ibid.

67. Van Keuren, "The NRL," 226.

68. Ibid.

69. Interview with Avery Richardson, 12 February 1991, and letter to author, 19 June 1991.

70. Quoted in Howeth, *Communications-Electronics*, 401.

71. NARA, RG 19, BuShips, Box 452. TWX from WPB, N.Y., to BuShips, 10 March 1945.

72. Deloraine, *Telecom*, 142. Deloraine's recollection of these events differs from the implication of the WPB telegram. Writing about the ITT facilities at

67 Broad Street, Deloraine noted that "By 1943 additional space was clearly necessary again, and the U.S. Navy agreed to support a plan for a building in Nutley, New Jersey." Once again this may have been a case of "the Navy" speaking with several different voices.

73. Kinert, "Naval Wartime Communication," 193.

74. Dow, "Navy Radio and Electronics," 287.

75. Howeth, *Communications-Electronics,* 436, refers to war as "expensive and wasteful" when describing the contracts that were canceled after the successful Normandy landings, without compensation to the manufacturers.

76. Busignies, Adams, and Colin, "Aerial Navigation," 113.

77. Howeth, *Communications-Electronics,* 402, 422, 546.

78. NHC/OA, Tenth Flt, ASM, Box 6, 1942. Unsigned and undated memo, "Recommendations in regard to H/F D/F Training Program," 2. "It is also necessary that a more ambitious program of training be undertaken in order that we may have trained operators available to man our own 500 sets of H/F D/F equipment when delivery starts about July, 1943."

Chapter 8   The Operators: Installation and Training

1. NARA, RG 19, BuShips, Gen. Corr., Box 1060. Samuelson to Navy Dept., 17 May 1940.

2. Interview with Avery Richardson, 19 February 1991.

3. Kinert, "Naval Wartime Communication," 193.

4. O'Connell, Pachynski, and Howeth, "Military Communication," 1246.

5. NRL Report No. R-2229, 5 February 1944, 2.

6. Gannon, *Operation Drumbeat,* 384.

7. O'Connell, Pachynski, and Howeth, "Military Communication," 1246.

8. NARA, RG 19, BuShips, Gen. Corr., Box 1062. VCNO to CINCPAC and CINCLANT, 28 August 1942.

9. NHC/OA, Tenth Flt, ASM, Box 6. HF/DF memo, 11 August 1942.

10. Ibid.

11. Ibid.

12. Keen, *Wireless Direction Finding,* 510–11.

13. NHC/OA, Tenth Flt, ASM, Box 6. HF/DF memo, 11 August 1942.

14. Ibid.

15. Ibid.

16. Ibid.

17. NHC/OA, Action Report, Box 109, No. 0023. TG 22.3, 22 June 1944.

18. NHC/OA, Tenth Flt, ASM, Box 6. HF/DF memo, 11 August 1942.

19. Rohwer and Hummelchen, *Chronology,* 139.

20. Sternhell and Thorndike, OEG, 41, notes that HF/DF came into operation in the fall of 1942, and that "by November 1942 HF/DF was accepted as an essential part of the equipment of the escort craft."

21. NRL Report No. R-1896, 20 June 1942, 16.

22. NHC/OA, Tenth Flt, ASM, Box 6. Comdr. TF 24 to CINCLANT, 27 October 1942.

23. Ibid.

24. NHC/OA, Tenth Flt, ASM, Box 6. CINCLANT to Commander Destroyers, Atlantic Fleet, 11 November 1942.

25. NHC/OA, Tenth Flt, ASM, Box 6, 1942. Unsigned memo, "Allocation HF/DF," cites letter of 18 November 1942 from VCNO.

26. NHC/OA, Tenth Flt, ASM, Box 6, 1942. Unsigned and undated memo, "Recommendations in regard to H/F D/F Training Program."

27. Ibid.

28. Ibid.

29. Robert Seroskie, letter to author, 12 May 1992.

30. NHC/OA Tenth Flt, ASM, Box 6. HF/DF Course, 28 February 1943.

31. Ibid.

32. Rohwer and Hummelchen, *Chronology*, 185. See also MacIntyre, *Battle of the Atlantic*, 150; and Milner, *U-Boat Hunters*, 29.

33. Robert Seroskie letters of 26 March and 12 May 1992. For confirmation see Rohwer and Hummelchen, *Chronology*, 215, 216, 222, 231.

34. NHC/OA, Tenth Flt, ASM, Box 6, 1942. "Allocation HF/DF," unsigned memo.

35. Interview with Avery Richardson, 19 February 1991.

36. Morison, vol. 10, *Atlantic Battle Won*, 77. See also Captain Brooks letter to author, 3 September 1993; and Rohwer and Hummelchen, *Chronology*, 221, 231, 235, 238, 251, for the *Belknap*'s operations.

37. Interview with Captain Brooks, 9 April 1992. See also Rohwer and Hummelchen, *Chronology*, 327. The *Belknap* ended up an APD (Auxiliary Personnel Destroyer) fast transport in the Pacific in January 1945, where it was damaged by kamikaze attack and was never repaired.

38. NHC/OA, Tenth Flt, ASM, Box 12. VCNO to Commander Destroyers Pacific Fleet, 14 September 1943.

39. Robert Seroskie letter to author, 12 May 1992.

40. Captain Brooks, letter to Captain Scott Lothrop, 31 January 1992.

41. NHC/OA, Tenth Flt, ASM, Box 12. VCNO to Commander Destroyers Pacific Fleet, 14 September 1943.

42. NHC/OA, Double Zero Files, Box 56. Report on "Measures for Combatting the Submarine Menace," report by Combined Staff Planners, 1 March 1943, 124.

43. NHC/OA, Action Report, Box 1240, No. 0026. TG 22.1, 25 November 1944.

44. Interview with Avery Richardson, 12 February 1991.

45. Ibid.

46. NHC/OA, Double Zero Files, Box 56. Report on "Measures for Combatting the Submarine Menace," report by Combined Staff Planners, 1 March 1943, Annex A to Appendix F, "U.S. Anti-Submarine Organization."

47. Ibid.

48. Captain Brooks, letter to Capt. Scott Lothrop, 21 February 1992.

49. Rohwer and Hummelchen, *Chronology*, 139.

50. Ibid., 170, 199.

51. Ibid., 345. See also Lundeberg, "Operation Teardrop Revisited," 210–30.

52. NHC/OA, Action Report, Box 107, No. 0044. TG 22.1, 27 April 1945.

53. Ibid.

54. Ibid.

55. Sternhell and Thorndike, OEG, 58. Howeth, *Communications-Electronics*, 545, erroneously claims that by September-October 1944 the U-boats had ceased using high-frequency radio altogether.

56. NHC/OA, Action Report, Box 107, No. 0044. TG 22.1, 27 April 1945.

57. Ibid.

58. Ibid.

59. NHC/OA, Action Report, Box 896 (no serial no.). TG 21.14, 10 September 1943.

60. Robert Seroskie, letter to author, 12 May 1992.

61. NHC/OA, Action Report, Box 107, No. 0012. TG 21.16, 6 April 1944.

62. NHC/OA, Action Report, Box 107, No. 0044. TG 22.1, 27 April 1945.

63. Ibid.

64. Lundeberg, "Operation Teardrop Revisited," 210–30.

65. NHC/OA, Tenth Flt, ASM, Box 6, 1942. Undated, unsigned memo, "British Specs for HF/DF Officer."

66. Captain Brooks, letter to Capt. Scott Lothrop, 31 January 1992.

67. Ibid.

68. NHC/OA, Action Report, Box 1240, No. 0026. TG 22.1, 25 November 1944.

69. Quoted in Rohwer and Douglas, "Canada and the Wolf Packs," 163.

Chapter 9 The Action: Huff Duff at War

1. Milner, *U-Boat Hunters*, 21–26.

2. Rohwer and Hummelchen, *Chronology*, 187–88. See also Milner, *U-Boat Hunters*, 26–30.

3. Kahn, *Seizing the Enigma,* 147.

4. Milner, *U-Boat Hunters,* 260–61. See also Rohwer, *Critical Convoy Battles,* 187–88.

5. MacIntyre, *Battle of the Atlantic,* 193.

6. Dönitz, *Memoirs,* 339.

7. MacIntyre, *Battle of the Atlantic,* 191–92. See also Syrett, *Defeat of German U-Boats,* 116.

8. Syrett, *Defeat of German U-Boats,* 134–41.

9. NHC/OA, Action Report, Box 855, No. 026. 6th SG, 29 May 1943.

10. Ibid.

11. Ibid. See also Y'Blood, *Hunter-Killer,* 282.

12. Morison, vol. 10, *Atlantic Battle Won,* 370.

13. NHC/OA, Action Report, Box 855, No. 026. 6th SG, 29 May 1943.

14. NHC/OA, Double Zero Files, Box 56. "Measures for Combatting the Submarine Menace," report by Combined Staff Planners, 1 March 1943, 59.

15. Interview with Captain Brooks, 9 April 1992.

16. Hezlet, *Electronics,* 236.

17. NHC/OA, Action Report, Box 104, No. 002. TG 21.12, 21 June 1943.

18. Ibid.

19. Ibid.

20. Ibid. See also Rohwer and Hummelchen, *Chronology,* 215.

21. NHC/OA, Action Report, Box 104, No. 002. TG 21.12, 21 June 1943. Captain Short reports a probable sinking on 4 June. Rohwer and Hummelchen, *Chronology,* 215, notes sinking of U-217 on 5 June.

22. NHC/OA, Action Report, Box 104, No. 002. TG 21.12, 21 June 1943.

23. Ibid.

24. Morison, vol. 10, *Atlantic Battle Won,* 112–13.

25. NHC/OA, Action Report, Box 104, No. 002. TG 21.12, 21 June 1943.

26. Ibid.

27. Ibid.

28. NHC/OA, Action Report, Box 896 (no serial no.). TG 21.14, 10 September 1943. See also Rohwer and Hummelchen, *Chronology,* 221.

29. NHC/OA, Action Report, Box 896 (no serial no.). TG 21.14, 10 September 1943.

30. Y'Blood, *Hunter-Killer,* 282. See also Rohwer and Hummelchen, *Chronology,* 221.

31. NHC/OA, Action Report, Box 896 (no serial no.). TG 21.14, 10 September 1943.

32. Ibid.

33. Interview with Captain Brooks, 9 April 1992.

34. NHC/OA, Action Report, Box 896 (no serial no.). TG 21.14, 10 September 1943.

35. Ibid.

36. Ibid. Four-stackers were the old World War I destroyers named for their four distinctive smokestacks.

37. Sternhell and Thorndike, OEG, 40.

38. Rohwer and Hummelchen, *Chronology,* 221.

39. NHC/OA, Action Report, Box 896 (no serial no.). TG 21.14, 10 September 1943.

40. Ibid.

41. Ibid.

42. NHC/OA, Action Report, Box 105, No. 0043. TG 21.14, 9 November 1943.

43. Ibid. The resume of operations reports that the *Borie* sank "one maybe two subs." See also Y'Blood, *Hunter-Killer,* 282; and Rohwer and Hummelchen, *Chronology,* 238.

44. NHC/OA, Action Report, Box 105, No. 0043. TG 21.14, 9 November 1943.

45. Rohwer and Douglas, "Canada and the Wolf Packs," 168.

46. There is some discrepancy in escort-group designations and vessels. See Syrett, *Defeat of German U-Boats,* 186; Rohwer and Hummelchen, *Chronology,* 236; and Rohwer and Douglas, "Canada and the Wolf Packs," 168–69.

47. Rohwer and Hummelchen, *Chronology,* 236. See also Syrett, *Defeat of German U-Boats,* 268–69, 272.

48. Rohwer and Douglas, "Canada and the Wolf Packs," 181–82. See also Syrett, *Defeat of German U-Boats,* 268–69.

49. Syrett, *Defeat of German U-Boats,* 257–58.

50. NHC/OA, Action Report, Box 105, No. 003. TG 21.14, 2 January 1944. Y'Blood, *Hunter-Killer,* 283, confirms sinking of U-645.

51. NHC/OA, Action Report, Box 105, No. 003. TG 21.14, 2 January 1944.

52. Ibid.

53. Ibid.

54. Morison, vol. 10, *Atlantic Battle Won,* 174–75.

55. NHC/OA, Action Report, Box 105, No. 003. TG 21.14, 2 January 1944.

56. Ibid.

57. NHC/OA, Action Report, Box 107, No. 0012. TG 21.16, 10 March 1944.

58. Ibid.

59. Ibid.

60. Y'Blood, *Hunter-Killer,* 283. See also Rohwer and Hummelchen, *Chronology,* 263.

61. NHC/OA, Action Report, Box 938, No. 0001. TG 21.15, 11 May 1944.

62. Ibid.

63. Ibid.

64. NHC/OA, Action Report, Box 938, No. 0001. TG 21.15 Study of Operations, 24 March to 11 May 1944, by P. R. Heineman, 22 May 1944.

65. Robert Seroskie, letter to author, 12 May 1992.

66. NHC/OA, Action Report, Box 103, No. 0025. TG 21.11, 29 May 1944.

67. E. J. Kahn, "Hand to Hand," 11–15.

68. Karl Degener-Böning, letter to author, 10 February 1991.

69. Ibid.

70. NHC/OA, Action Report, Box 856, No. 0035. TG 22.3, 24 September 1944. See also Rohwer and Hummelchen, *Chronology*, 289.

71. NHC/OA, Action Report, Box 856, No. 0035. TG 22.3, 24 September 1944.

72. NHC/OA, Action Report, Box 1245, No. 075. Report from the CO of the USS *Moffett* on engagement with an enemy sub, 17 May 1943. The original action report sheet, with its interesting heading, is in the file. Perhaps this is only equaled by Dönitz's directive to his U-boats in September 1943, "Fire first, then submerge." Quoted in Rohwer and Douglas, "Canada and the Wolf Packs," 168.

73. Tidman, *The OEG*, 63.

74. NHC/OA, Ready Reference File, HF/DF.

75. Y'Blood, *Hunter-Killer*, 282–83.

76. Sternhell and Thorndike, OEG, 80, estimates that approximately 733 German submarines were sunk by the Allies in the course of the war. Prof. Jürgen Rohwer, letter to author of 8 December 1994, puts the number of U-boats lost at 630, plus an additional 81 lost in harbor and in home waters from air attacks and mines. Forty-two were also lost by accidents in harbor and home waters.

77. Cross report, vol. 1, ix.

Chapter 10   *The Verdict on Huff Duff: World War II and After*

1. Prof. Jürgen Rohwer, letter to author, 8 December 1994. Hessler's numbers are close; see *U-Boat War*, vol. 3, 101.

2. van der Vat, *Atlantic Campaign*, 382. See also Hessler, *U-Boat War*, vol. 3, 101, for his estimate of Allied shipping losses.

3. United States Fiftieth Anniversary of World War II Commemoration Committee: General Fact Sheet of WWII.

4. van der Vat, *Atlantic Campaign*, 382.

5. Ibid.

6. NRL war history, 82.

7. *New York Times*, 14 January 1946, 1.

8. Ibid.

9. Ibid.

10. See articles appearing on 14 January 1946 in the *Detroit News*, 3; the *St. Louis Post-Dispatch*, 1; the *New Orleans Times Picayune*, 21; the *Baltimore Morning Sun*, 1, 5; and the *Washington Post*, 11.

11. *New York Times*, 8.

12. Ibid.

13. Ibid.

14. Gebhard, *Radio-Electronics*, 307.

15. Cross report, vol. 1, 156. Testimony of Capt. Norval Richardson, 189–90.

16. Cross report, vol. 1, 2–33.

17. Morton, "Can We Manage the Radio-Electronic Battle?" 92.

18. Ibid.

19. Ibid.

20. Ibid.

21. Ibid.

22. Interview with Captain Brooks, 9 April 1992.

Bibliography

Author's Interviews and Correspondence

Brooks, Capt. Benjamin T., USN (Ret.). Skipper of the *Belknap*.
Busignies, Mme Cécile. Widow of Henri Busignies.
Degener-Böning, Karl. Radio operator, U-66.
Furth, Rear Adm. Frederick R., USN (Ret.). Technical adviser to Rear Adm. Leigh Noyes, Dir. Naval Communications.
Hirschfeld, Wolfgang. Radio operator, U-109, U-234.
Koz Paley, Barbara Goldstein. Daughter of Dr. M. K. Goldstein.
Lombardi, Eugene. Engineer with ITT.
Lombardi, Ruth Lockhart. Secretary and technical assistant to Henri Busignies.
Lothrop, Capt. Scott, USN (Ret.). USS *Tarbell*.
Richardson, Avery G. Engineer with ITT.
Rohwer, Dr. Jürgen.
Seroskie, Robert J. Radioman and HF/DF operator.
Sivowitch, Elliot N. Museum specialist, National Museum of American History, Division of Electricity and Modern Physics, Smithsonian Institution.
Van Keuren, Dr. David K. Head of History Department, NRL.

Unpublished Sources

Henri Busignies Papers (in possession of Mme Busignies)
Biographical sketch, April 1973 (typewritten).
"Lectures and Papers by Dr. Busignies," 1975.
List of "Most Important Patents" (handwritten).
List of "Publications by Dr. Busignies."
Memoir of Henri Busignies's career in electronics.

Maxwell K. Goldstein Papers (in possession of Mrs. B. Goldstein Koz Paley)
Distinguished civilian service award citation.
Johns Hopkins University class of 1930 yearbook.

Two NRL letters concerning draft status.
Papers relating to patent application No. 409,391, for a "Phase Type Direct Indicating Direction Finder."
"Resume of Professional Background" (typewritten).
"Some Research and Development Experiences in the U.S. Navy" (typewritten memoir).
"Vita of Dr. M. K. Goldstein, Naval Research Laboratory" (typewritten).

N<small>ATIONAL</small> A<small>RCHIVES AND</small> R<small>ECORDS</small> A<small>DMINISTRATION</small>

Record Group 19, Bureau of Ships: Boxes 452, 617, 1060, 1061, 1062, 1063, 1070, 1070A.
Record Group 74, Bureau of Ships: Box 99.

N<small>AVAL</small> H<small>ISTORICAL</small> C<small>ENTER</small>, N<small>AVY</small> D<small>EPARTMENT</small> L<small>IBRARY</small>

U.S. Navy, Bureau of Ships, Naval Research Laboratory, *World War II Administrative History*, No. 134. "War History of the Naval Research Laboratory." 1 November 1946.

N<small>AVAL</small> H<small>ISTORICAL</small> C<small>ENTER</small>, O<small>PERATIONAL</small> A<small>RCHIVES</small>

Action Reports: Box 103, Serial Number 0025; Box 104, No. 002; Box 105, Nos. 003, 0043; Box 107, Nos. 0012, 0044; Box 109, No. 0023; Box 171, Nos. 083, 090; Box 855, No. 026; Box 856, No. 050, 0035, 0045; Box 896; Box 938, Nos. 0001, 0020; Box 1240, No. 0026; Box 1245, No. 075.
Analysis and Statistics Section Files, Boxes 43, 44.
Double Zero Files: Boxes 2, 41, 42, 49, 56.
King Papers: Box 13.
Ready Reference Files: HF/DF; Ad. E. M. Eller Letters
Tenth Fleet: Anti-Submarine Measures Division Files, Boxes 6, 12, 13, 43, 45, 47.

N<small>AVAL</small> R<small>ESEARCH</small> L<small>ABORATORY</small>

Report No. R-1896. "The Performance of the British FH-3 and the International Telephone and Radio Laboratory's Crossed Loop Type High Frequency Direction Finders." 20 June 1942.
Report No. R-2229. "Interpolation Charts for HF-DF Shipboard Calibration." 5 February 1944.

O<small>THER</small> U<small>NPUBLISHED</small> W<small>ORKS</small>

Deloraine, Maurice. "L'Histoire du Huff-Duff" (typewritten). Undated, in possession of Mme Busignies.

Plante, George. "A Very Personal View of World War II." 1991 (typewritten memoir). In possession of author.

Published Sources

BOOKS

Abbazia, Patrick. *Mr. Roosevelt's Navy: The Private War of the United States Atlantic Fleet, 1939–1942.* Annapolis, Md.: Naval Institute Press, 1977.

Allard, Dean C. "A United States Overview." In *The Battle of the Atlantic, 1939–1945,* ed. Stephen Howarth and Derek Law. Annapolis, Md.: Naval Institute Press, 1994.

Allison, David Kite. *New Eye for the Navy: The Origin of Radar at the Naval Research Laboratory.* Washington, D.C.: Naval Research Laboratory, 1981.

Auphan, Amiral, and Jacques Mordal. *La Marine française dans la Seconde Guerre Mondiale.* Paris: Editions France-Empire, 1967.

Barnett, Correlli. *Engage the Enemy More Closely: The Royal Navy in the Second World War.* New York: W. W. Norton, 1991.

Beesly, Patrick. *Very Special Intelligence: The Story of the Admiralty's Operational Intelligence Centre, 1939–1945.* London: Hamish Hamilton, 1977.

Bond, Brian. *War and Society in Europe.* New York: Oxford University Press, 1986.

Boutilier, James A., ed. *The RCN in Retrospect, 1910–1968.* Vancouver: University of British Columbia Press, 1982.

Brown, David K. "Atlantic Escorts, 1939–1945." In *The Battle of the Atlantic, 1939–1945,* ed. Stephen Howarth and Derek Law. Annapolis, Md.: Naval Institute Press, 1994.

Churchill, Winston S. *The Second World War.* Vol. 2, *Their Finest Hour.* Boston: Houghton Mifflin, 1949.

Connery, Robert H. *The Navy and Industrial Mobilization in World War II.* Princeton: Princeton University Press, 1951.

Cremer, Peter. *U-Boat Commander: A Periscope View of the Battle of the Atlantic.* Trans. by Lawrence Wilson. Annapolis, Md.: Naval Institute Press, 1984.

Deloraine, Maurice. *When Telecom and ITT Were Young.* New York: Lehigh Books, 1976.

Dönitz, Karl. *Memoirs: Ten Years and Twenty Days.* Trans. by R. H. Stevens, with new introduction and afterword by Jürgen Rohwer. Annapolis, Md.: Naval Institute Press, 1990.

Douglas, W.A.B. *The RCN in Transition, 1910–1985.* Vancouver: University of British Columbia Press, 1988.

Douglas, W.A.B., and Jürgen Rohwer. "The Most Thankless Task Revisited: Convoys, Escorts, and Radio Intelligence in the Western Atlantic, 1941–1943." In *The RCN in Retrospect, 1910–1968,* ed. James A. Boutilier. Vancouver: University of British Columbia Press, 1982.

Elliott, Peter. *Allied Escort Ships of World War II.* London: Macdonald and Jane's, 1977.

Farago, Ladislas. *The Tenth Fleet.* New York: Ivan Obolensky, Inc., 1962.

Fisher, David E. *A Race on the Edge of Time: Radar, The Decisive Weapon of World War II.* New York: Paragon House, 1989.

Friedman, Norman. *Submarine Design and Development.* Annapolis, Md.: Naval Institute Press, 1984.

———. *The U.S. Maritime Strategy.* London: Jane's Publishing, 1988.

Führer Conferences on Naval Affairs, 1939–1945. Foreword by Jak P. Mallman Showell. Annapolis, Md.: Naval Institute Press, 1990.

Gannon, Michael. *Operation Drumbeat: The Dramatic True Story of Germany's First U-Boat Attacks Along the American Coast in World War II.* New York: Harper & Row, 1990.

Gardner, W.J.R. "An Allied Perspective." In *The Battle of the Atlantic, 1939–1945,* ed. Stephen Howarth and Derek Law. Annapolis, Md.: Naval Institute Press, 1994.

Gebhard, Louis A. *Evolution of Naval Radio-Electronics and Contributions of the Naval Research Laboratory.* NRL Report No. 8300. Washington, D.C.: Naval Research Laboratory, 1979.

Gretton, Vice Adm. Sir Peter. *Crisis Convoy: The Story of HX 231.* Annapolis, Md.: Naval Institute Press, 1974.

Guske, Heinz F. K. *The War Diaries of U-764.* Gettysburg, Pa.: Thomas Publications, 1992.

Hackmann, Willem. *Seek and Strike: Sonar, Anti-Submarine Warfare and the Royal Navy, 1914–1954.* London: Her Majesty's Stationery Office, 1984.

Hague, Arnold. *The Towns: A History of the Fifty Destroyers Transferred From the United States to Great Britain in 1940.* Kendal, England: World Ship Society, 1988.

Hessler, Günter. *The U-Boat War in the Atlantic, 1939–1945.* A facsimile edition in 1 vol. of the 3 vols. prepared by G. Hessler for the Admiralty, London: Her Majesty's Stationery Office, 1989.

Hezlet, Vice Adm. Sir Arthur. *Electronics and Sea Power.* New York: Stein & Day, 1975.

Hinsley, F. H., et al. *British Intelligence in the Second World War,* abridged edition. Cambridge: Cambridge University Press, 1993.

Hirschfeld, Wolfgang. *Feindfahrten: Das Logbuch Eines U-Bootfunkers.* N.p.: Neuer Kaiser Verlag, 1991.

Holmes, W. J. *Double-Edged Secrets.* Annapolis, Md.: Naval Institute Press, 1977.

Hoopes, Townsend, and Douglas Brinkley. *Driven Patriot: The Life and Times of James Forrestal.* New York: Knopf, 1992.

Howarth, Stephen, and Derek Law, eds. *The Battle of the Atlantic, 1939–1945.* Annapolis, Md.: Naval Institute Press, 1994.

Howeth, Capt. L. S. *History of Communications-Electronics in the United States Navy.* Washington, D.C.: Bureau of Ships and Office of Naval History, 1963.

Howse, Derek. *Radar at Sea: The Royal Navy in World War 2.* Annapolis, Md.: Naval Institute Press, 1993.

Hyde, H. Montgomery. *Room 3603: The Story of the British Intelligence Center in New York During World War II.* New York: Farrar, Straus, 1963.

Jones, R. V. *Most Secret War.* New York: Coronet Books, 1979.

Kahn, David. *Seizing the Enigma: The Race to Break the German U-Boat Codes, 1939–1943.* Boston: Houghton Mifflin, 1991.

Keegan, John. *The Mask of Command.* New York: Penguin, 1988.

———. *The Price of Admiralty.* New York: Penguin, 1990.

Keen, R. *Wireless Direction Finding.* London: Iliffe & Sons, 1947.

King, Adm. Ernest J. "Cominch Takes a Hard Look at the U-Boat Situation." Chap. 6 in *The U.S. Navy in World War II,* ed. S. E. Smith. New York: William Morrow, 1966.

Love, Robert W., Jr. *History of the U.S. Navy,* vol. 2. Harrisburg, Pa.: Stackpole Books, 1992.

———. "The U.S. Navy and Operation Roll of Drums, 1942." In *To Die Gallantly: The Battle of the Atlantic,* ed. Timothy J. Runyan and Jan M. Copes. Boulder, Colo.: Westview Press, 1994.

Lundeberg, Philip. "Allied Cooperation." In *The Battle of the Atlantic, 1939–1945,* ed. Stephen Howarth and Derek Law. Annapolis, Md.: Naval Institute Press, 1994.

————. "Operation Teardrop Revisited." In *To Die Gallantly: The Battle of the Atlantic,* ed. Timothy J. Runyan and Jan M. Copes. Boulder, Colo.: Westview Press, 1994.

MacIntyre, Donald. *The Battle of the Atlantic.* London: B. T. Batsford, 1961.

Martin, James Stewart. *All Honorable Men.* Boston: Little, Brown, 1950.

McCue, Brian. *U-Boats in the Bay of Biscay: An Essay in Operations Analysis.* Washington, D.C.: National Defense University Press, 1990.

McKay, Ernest A. *Undersea Terror: U-boat Wolf-Packs in World War II.* New York: Julian Messner, 1982.

Meigs, Montgomery C. *Slide Rules and Submarines.* Washington, D.C.: National Defense University Press, 1990.

Middlebrook, Martin. *Convoy.* New York: Quill, 1976.

Milner, Marc. *North Atlantic Run: The Royal Canadian Navy and the Battle of the Convoys.* Toronto: University of Toronto Press, 1985.

————. *The U-Boat Hunters.* Annapolis, Md.: Naval Institute Press, 1994.

Morison, Samuel Eliot. *History of United States Naval Operations in World War II.* Vol. 1, *Battle of the Atlantic, 1939–1943,* and vol. 10, *Atlantic Battle Won, May 1943–May 1945.* Boston: Little, Brown, 1947–62.

Niestlé, Axel. "German Technical and Electronic Development." In *The Battle of the Atlantic, 1939–1945,* ed. Stephen Howarth and Derek Law. Annapolis, Md.: Naval Institute Press, 1994.

O'Neill, William L. *A Democracy at War: America's Fight at Home and Abroad in World War II.* New York: Free Press, 1993.

Padfield, Peter. *Dönitz: The Last Führer.* New York: Harper & Row, 1984.

Pugh, Philip. "Military Need and Civil Necessity." In *The Battle of the Atlantic, 1939–1945,* ed. Stephen Howarth and Derek Law. Annapolis, Md.: Naval Institute Press, 1994.

Rahn, Werner. "The Campaign: The German Perspective." In *The Battle of the Atlantic, 1939–1945,* ed. Stephen Howarth and Derek Law. Annapolis, Md.: Naval Institute Press, 1994.

Raven, Alan, and John Roberts. *British Battleships of World War II.* Annapolis, Md.: Naval Institute Press, 1976.

R. F. Cross Associates. *Sea-Based Airborne Antisubmarine Warfare, 1940–1977.* 3 vols. Alexandria, Va.: R. F. Cross Associates, 1978.

Rhys-Jones, Graham. "The German System: A Staff Perspective." In *The Battle of the Atlantic, 1939–1945,* ed. Stephen Howarth and Derek Law. Annapolis, Md.: Naval Institute Press, 1994.

Rohwer, Jürgen. "Die Auswirkungen der Deutschen und Britischen Funkaufklärung auf die Geleitzugoperationen im Nordatlantik." In *Die Funkaufklärung und Ihre Rolle im Zweiten Weltkrieg,* ed. Jürgen Rohwer and Eberhard Jäckel. Stuttgart: Motorbuch Verlag, 1979.

———. *Axis Submarine Successes,* English language ed. Annapolis, Md.: Naval Institute Press, 1983.

———. *The Critical Convoy Battles of March 1943: The Battle for HX.229/SC.122.* Annapolis, Md.: Naval Institute Press, 1977.

———. "The Wireless War." In *The Battle of the Atlantic, 1939–1945,* ed. Stephen Howarth and Derek Law. Annapolis, Md.: Naval Institute Press, 1994.

Rohwer, Jürgen, and G. Hummelchen. *Chronology of the War at Sea,* expanded edition. Annapolis, Md.: Naval Institute Press, 1992.

Rohwer, Jürgen, and W.A.B. Douglas. "Canada and the Wolf Packs, September 1943." In *The RCN in Transition, 1910–1985,* ed. W.A.B. Douglas. Vancouver: University of British Columbia Press, 1988.

Roscoe, Theodore. *United States Destroyer Operations in World War II.* Annapolis, Md.: Naval Institute Press, 1953.

Roskill, S. W. *The War at Sea, 1939–1945.* 3 vols. London: Her Majesty's Stationery Office, 1954–61.

Rössler, Eberhard. "U-Boat Development and Building." In *The Battle of the Atlantic, 1939–1945,* ed. Stephen Howarth and Derek Law. Annapolis, Md.: Naval Institute Press, 1994.

Runyan, Timothy J., and Jan M. Copes, eds. *To Die Gallantly: The Battle of the Atlantic.* Boulder, Colo.: Westview Press, 1994.

Sadkovich, James J. *The Italian Navy in World War II.* Westport, Conn.: Greenwood Press, 1994.

Salewski, Michael. "The Submarine War." In *U-Boat War,* by Lothar-Gunther Buchheim. New York: Bantam Books, 1979.

Sampson, Anthony. *The Sovereign State of ITT.* New York: Stein & Day, 1973.

Sarty, Roger, and Jürgen Rohwer. "Intelligence and the Air Forces in the Battle of the Atlantic, 1943–1945." In *Acta,* No. 13. Helsinki: International Commission of Military History, 1991.

Smith, Bradley F. *The Ultra-Magic Deals and the Most Secret Special Relationship, 1940–1946.* Novato, Calif.: Presidio Press, 1994.

Sobel, Robert. *ITT: The Management of Opportunity.* New York: Truman Talley Books, 1982.

Sternhell, Charles M., and Alan M. Thorndike. *Antisubmarine Warfare in World War II.* Operations Evaluation Group Report No. 51. Washington, D.C.: Office of the Chief of Naval Operations, Navy Department, 1946.

Syrett, David. *The Defeat of the German U-Boats: The Battle of the Atlantic.* Columbia: University of South Carolina Press, 1994.

Terraine, John. *The U-Boat Wars, 1916–1945.* New York: G. P. Putnam's Sons, 1989.

Tidman, Keith R. *The Operations Evaluation Group: A History of Naval Operations Analysis.* Annapolis, Md.: Naval Institute Press, 1984.

Till, Geoffrey. "The Battle of the Atlantic As History." In *The Battle of the Atlantic, 1939–1945,* ed. Stephen Howarth and Derek Law. Annapolis, Md.: Naval Institute Press, 1994.

Topp, Erich. "Manning and Training the U-Boat Fleet." In *The Battle of the Atlantic, 1939–1945,* ed. Stephen Howarth and Derek Law. Annapolis, Md.: Naval Institute Press, 1994.

Travers, Douglas N., and Stuart M. Hixon, eds. *Abstracts of the Available Literature on Radio Direction Finding, 1899–1965.* San Antonio, Tex.: Southwest Research Institute, 1966.

van der Vat, Dan. *The Atlantic Campaign: World War II's Great Struggle at Sea.* New York: Harper & Row, 1988.

Walton, Francis. *Miracle of World War II: How American Industry Made Victory Possible.* New York: Macmillan, 1956.

Watson-Watt, R. A. *Three Steps to Victory.* London: Odhams Press, 1957.

Werner, Herbert A. *Iron Coffins: A Personal Account of the German U-Boat Battles of World War II.* New York: Holt, Rinehart & Winston, 1969.

Winterbotham, F. W. *The Ultra Secret.* New York: Dell, 1974.

Y'Blood, William T. *Hunter-Killer: U.S. Escort Carriers in the Battle of the Atlantic.* Annapolis, Md.: Naval Institute Press, 1983.

Zimmerman, David. *The Great Naval Battle of Ottawa.* Toronto: University of Toronto Press, 1988.

———. "Technology and Tactics." In *The Battle of the Atlantic, 1939–*

1945, ed. Stephen Howarth and Derek Law. Annapolis, Md.: Naval Institute Press, 1994.

PERIODICALS

Adams, Paul R., and Robert I. Colin. "Frequency, Power and Modulation of a Long-Range Radio Navigation System." *Electrical Communication* 23 (June 1946): 144–58.

Bennett, Capt. R. "Electronics in Naval Warfare." *Proceedings: Institute of Radio Engineers* 34 (February 1946): 81.

Busignies, Henri. "The Automatic Radio Compass and Its Applications to Aerial Navigation." *Electrical Communication* 15 (October 1936): 157–72.

———. "Communication and Navigation." *Proceedings: Institute of Radio Engineers* 50 (1962): 592.

———. "Evaluation of Night Errors in Aircraft Direction Finding, 150–1500 Kilocycles." *Electrical Communication* 23 (March 1946): 42–62.

Busignies, Henri, Paul R. Adams, and Robert I. Colin. "Aerial Navigation and Traffic Control with Navaglobe, Navar, Navaglide, and Navascreen." *Electrical Communication* 23 (June 1946): 112–43.

"Contributors to This Issue." *Electrical Communication* 23 (March 1946): 109.

Deloraine, Maurice. "Telephone, Telegraph and Telex Traffic." *Electrical Communication* 39 (1964): 276–78.

De Walden, S., A.F.L. Rocke, J.O.G. Barrett, and W. J. Pitts. "The Development of a High-Frequency Cathode-Ray Direction Finder for Naval Use." *Journal of the Institute of Electrical Engineers* 94 (1947): 823–37.

Dow, Jennings B. "Navy Radio and Electronics During World War II." *Proceedings: Institute of Radio Engineers* 34 (May 1946): 284–87.

"Electrical Communication: 1940–1945, War Years' Review," in 3 parts. *Electrical Communication* 23 (March, June, September 1946): 3–18; 214–16; 339–66.

Erskine, Ralph. "U-Boats, Homing Signals and HFDF." *Intelligence and National Security* 2 (1987): 324–30.

Gething, P.J.D. "High-Frequency Direction-Finding." *Proceedings: Institute of Radio Engineers* 113 (January 1966): 49–50.

"Huff-Duff Presented for Smithsonian Display." *WITTS: Within the International Telephone and Telegraph System* (February 1976): 1

"Huff-Duff, U-Boat Finder, Seen as Defense Against Atomic Bomb." *New York Times.* 14 January 1946.

"Institute News and Radio Notes." *Proceedings: Institute of Radio Engineers* 34 (July 1946): 477.

Kahn, E. J., Jr. "Hand to Hand." *Naval History* 2 (Summer 1988): 11–15.

Kinert, J. O. "Naval Wartime Communication Problems." *Proceedings: Institute of Radio Engineers* 34 (April 1946): 193–95.

Morton, John F. "Can We Manage the Radio-Electronic Battle?" U.S. Naval Institute *Proceedings* 117 (January 1991): 92–96.

"Naval Research Laboratory: Its Contributions to War and Peace Are Reviewed." *Product Engineering* (June 1946): 492–96.

Norton, Kenneth A. "Radio-Wave Propagation during World War II." *Proceedings: Institute of Radio Engineers* 50 (May 1962): 698–704.

O'Connell, J. D., A. L. Pachynski, and L. S. Howeth. "Summary of Military Communication in the United States, 1860–1962." *Proceedings: Institute of Radio Engineers* 50 (May 1962): 1241–51.

Page, R. M. "Radar Model XAF." *Proceedings: Institute of Radio Engineers* 34 (February 1946): 88.

"Recent Telecommunications Developments." *Electrical Communication* 23 (March 1946): 241–43.

Richardson, Avery G. "From Galena to Gigahertz." Institute of Electrical and Electronic Engineers: *AES Magazine* (February 1988): 2–8.

"Spot Location by HF/DF Data." *Aero Digest* 52 (March 1946): 63

Syrett, David. "Communications Intelligence and the Battle of the Atlantic, 1943–1945." *Archives* 22 (April 1995): 45–59.

———. "The Safe and Timely Arrival of Convoy SC 130, 15–25 May 1943." *The American Neptune* 50 (Summer 1990): 219–27.

———. "Weather-Reporting U-Boats in the Atlantic, 1944–45: The Hunt for U-248." *The American Neptune* 51 (Winter 1992): 16–24.

Van Atta, L. C., and S. Silver. "Contributions to the Antenna Field During World War II." *Proceedings: Institute of Radio Engineers* 50 (May 1962): 692–97.

Van Keuren, Rear Adm. A. H. "The U.S. Naval Research Laboratory." *Journal of Applied Physics* 15 (March 1944): 221

Watson-Watt, R. A., and J. F. Herd. "An Instantaneous Direct-Reading Radiogoniometer." *Journal of the Institute of Electrical Engineers* 64 (1926): 611.

"The Winter Meeting." *Proceedings: Institute of Radio Engineers* 34 (March 1946): 37.

Index

Van Keuren, Rear Adm. A. H., 143, 159–
60, 180–81
Vessey, Gen. John, 239
Vest, Capt. John P. W., 224
Vichy France, 116–17, 118, 123–24, 131
Vosseller, Capt. Aurelius B., 226

Wake Island, USS (escort carrier), 226
War Production Board (WPB), 167, 169,
176, 181–82, 235
Washington, D.C., 78, 87, 119, 123, 132,
135, 151, 177, 216; convoy movements
directed from, 207; HF/DF information
sent to, 88; HF/DF net-plotting center
at, 198
Washington Convoy Conference. *See* Atlantic Convoy Conference

Watson-Watt, Sir Robert: and D/F, 81–82;
and distrust of Deloraine and France,
111–13; and HF/DF, 90, 103; and radar,
65–66, 81, 87; writes article on radiogoniometer, 81. *See also* radar
Willis, USS (destroyer escort), 227
Whimbrel, HMS (frigate), 199
WPB. *See* War Production Board

Y-stations. *See* radio receiving stations

Zaunkönig (acoustic torpedo), 52, 219,
220, 221
Z-plan, 22

About the Author

Kathleen Broome Williams was born in Charlottesville, Virginia, in 1944. Her father, Maj. Roger G. B. Broome, USMCR, died during the American landings on Saipan when she was four months old. She grew up in Italy and England, returning to the United States to attend Wellesley College.

Williams holds an M.A. from Columbia University and a Ph.D. in military history from the City University of New York. She is a compiler of *The Commissioned Sea Officers of the Royal Navy, 1660–1815*, and a contributor to *The Civil War Book of Lists*. Currently, she teaches at the John Jay College of Criminal Justice, CUNY, and is conducting research into the role of women in the development of naval technology.

The **Naval Institute Press** is the book-publishing arm of the U.S. Naval Institute, a private, nonprofit, membership society for sea service professionals and others who share an interest in naval and maritime affairs. Established in 1873 at the U.S. Naval Academy in Annapolis, Maryland, where its offices remain today, the Naval Institute has almost 85,000 members worldwide.

Members of the Naval Institute support the education programs of the society and receive the influential monthly magazine *Proceedings* and discounts on fine nautical prints and on ship and aircraft photos. They also have access to the transcripts of the Institute's Oral History Program and get discounted admission to any of the Institute-sponsored seminars offered around the country.

The Naval Institute also publishes *Naval History* magazine. This colorful bimonthly is filled with entertaining and thought-provoking articles, first-person reminiscences, and dramatic art and photography. Members receive a discount on *Naval History* subscriptions.

The Naval Institute's book-publishing program, begun in 1898 with basic guides to naval practices, has broadened its scope in recent years to include books of more general interest. Now the Naval Institute Press publishes about 100 titles each year, ranging from how-to books on boating and navigation to battle histories, biographies, ship and aircraft guides, and novels. Institute members receive discounts of 20 to 50 percent on the Press's nearly 600 books in print.

Full-time students are eligible for special half-price membership rates. Life memberships are also available.

For a free catalog describing Naval Institute Press books currently available, and for further information about subscribing to *Naval History* magazine or about joining the U.S. Naval Institute, please write to:

Membership Department
U.S. Naval Institute
118 Maryland Avenue
Annapolis, Maryland 21402-5035
Telephone: (800) 233-8764
Fax: (410) 269-7940